Sweet Biochemistry

Remembering Structures, Cycles, and Pathways by Mnemonics

Sweet Biochemistry

Remembering Structures, Cycles, and Pathways by Mnemonics

Asha Kumari
Department of Biochemistry, Pandit Bhagwat Dayal Sharma PGIMS,
Rohtak, Haryana, India

ELSEVIER
ACADEMIC PRESS
An imprint of Elsevier

Academic Press is an imprint of Elsevier
125 London Wall, London EC2Y 5AS, United Kingdom
525 B Street, Suite 1800, San Diego, CA 92101-4495, United States
50 Hampshire Street, 5th Floor, Cambridge, MA 02139, United States
The Boulevard, Langford Lane, Kidlington, Oxford OX5 1GB, United Kingdom

Notices
Knowledge and best practice in this field are constantly changing. As new research and experience broaden our understanding, changes in research methods, professional practices, or medical treatment may become necessary.

Practitioners and researchers must always rely on their own experience and knowledge in evaluating and using any information, methods, compounds, or experiments described herein. In using such information or methods they should be mindful of their own safety and the safety of others, including parties for whom they have a professional responsibility.

To the fullest extent of the law, neither the Publisher nor the authors, contributors, or editors, assume any liability for any injury and/or damage to persons or property as a matter of products liability, negligence or otherwise, or from any use or operation of any methods, products, instructions, or ideas contained in the material herein.

Library of Congress Cataloging-in-Publication Data
A catalog record for this book is available from the Library of Congress

British Library Cataloguing-in-Publication Data
A catalogue record for this book is available from the British Library

ISBN: 978-0-12-814453-4

For Information on all Academic Press publications
visit our website at https://www.elsevier.com/books-and-journals

Publisher: Sara Tenney
Acquisition Editor: Linda Versteeg-Buschman
Editorial Project Manager: Samuel Young
Production Project Manager: Anusha Sambamoorthy
Cover Designer: MPS

Typeset by MPS Limited, Chennai, India

DEDICATION

Dedicated to Lord Hanuman

Manojavam maruttulya vegam,

Jitendriyam budhdhimatam varishtham

vatatmajam vanaryudh mukhyam,

Shri Ram dutam sharanam prapadye

CONTENTS

LIST OF FIGURES

LIST OF TABLES

PREFACE

In our nursery classes, we all remember how we use to learn, for example, "A for Apple," "B for Ball." We used to see pictures, sing rhymes, and play with letters. As we grow older we feel that books for adults are monotonous, full of concentrated material, and that some act as sleeping pills for most of us. We do not teach a kid cram A B C D like genetic code sequence. We try to attach alphabets to objects which we can see and correlate. Study should be fun for everyone not only kids. We store our memories as visual images. Therefore, I have tried to make some of the intricate topics of biochemistry funny and memorable. The topics are presented as rhymes, picture stories, and formulae which will make you smile. You will meet syndromes represented as people.

This book is for all biochemistry lovers, including undergraduate and postgraduate students, paramedical students, and teaching faculty. During my teaching experience I have often seen students complaining that biochemistry is difficult to remember. I hope they will enjoy this presentation of structures and pathways.

In this book, for every topic a traditional short text is revised, followed by a new presentation. New mnemonics are described and correlated with the text. This book is not a textbook but an attempt to make the subject enjoyable.

I am eagerly awaiting feedback from my readers.

Asha Kumari

ACKNOWLEDGMENTS

This book has given me a chance to express gratitude to my parents, Sh. Dayanand and Smt. Burfo Devi, and affection to my brothers, sister-in-laws, and kids. I would like here to mention my guide, Dr. Shashi Seth, my favorite teacher in demonstrating me the values of life. I am also thankful to my Head of Department Dr. Veena Singh Ghalaut for her guidance in studies. My friends who constantly support and encourage my stupidities also deserve a place here.

Asha Kumari

CHAPTER 1

Glycolysis

TRADITIONAL GLYCOLYSIS RECAP

Glycolysis is a cytoplasmic pathway which breaks down glucose into two three-carbon compounds and generates energy. Glycolysis is used by all cells in the body for energy generation. Glucose is trapped by phosphorylation, with the help of the enzyme hexokinase. Adenosine triphosphate (ATP) is used in this reaction and the product, glucose-6-Phosphate (G-6-P), inhibits hexokinase. This is an irreversible reaction. G-6-P is isomerized into its ketose form, fructose-6-Phosphate (F-6-P), by phosphohexose isomerase.

F-6-P is further phosphorylated by phosphofructokinase to fructose 1,6-bisphosphate (F1,6 bisP). This reaction is irreversible and is the principal regulatory step. Aldolase cleaves F1,6 bisP into glyceraldehyde-3-P and dihydroxyacetone phosphate (DHAP), which are interconverted by the enzyme phosphotriose isomerase.

Glyceraldehyde-3-P is oxidized by NAD+-dependent dehydrogenase forming 1,3-bisphosphoglycerate.

1,3-Bisphosphoglycerate has a high-energy acyl phosphate bond and carries out substrate-level phosphorylation generating ATP. The participating enzyme is phosphoglycerate kinase and 3-phosphoglycerate is formed. Phosphoglycerate mutase isomerizes 3-phosphoglycerate to 2-phosphoglycerate. 2-Phosphoglycerate is dehydrated by enolase to form phosphoenolpyruvate, the second compound capable of substrate-level phosphorylation in glycolysis. Pyruvate kinase transfers the phosphate group of phosphoenolpyruvate to adenosine diphosphate (ADP) and pyruvate is formed.

Pyruvate enters the Krebs cycle in aerobic conditions, and in anaerobic conditions it forms lactate which helps in the generation of

Sweet Biochemistry: Remembering Structures, Cycles, and Pathways by Mnemonics
DOI: http://dx.doi.org/10.1016/B978-0-12-814453-4.00001-7

NAD+ for the continuation of glycolysis. Pyruvate is converted to acetyl-CoA by pyruvate dehydrogenase complex, which is an irreversible step. Pyruvate enters the Krebs cycle for further energy production.

Glycolysis is one of the basic metabolic pathways, and is crucial for the life of most organisms. We start this book with the pathway showing the reactions, substrates, products, enzymes, and other involved molecules of glycolysis (Figs. 1.1 and 1.2).

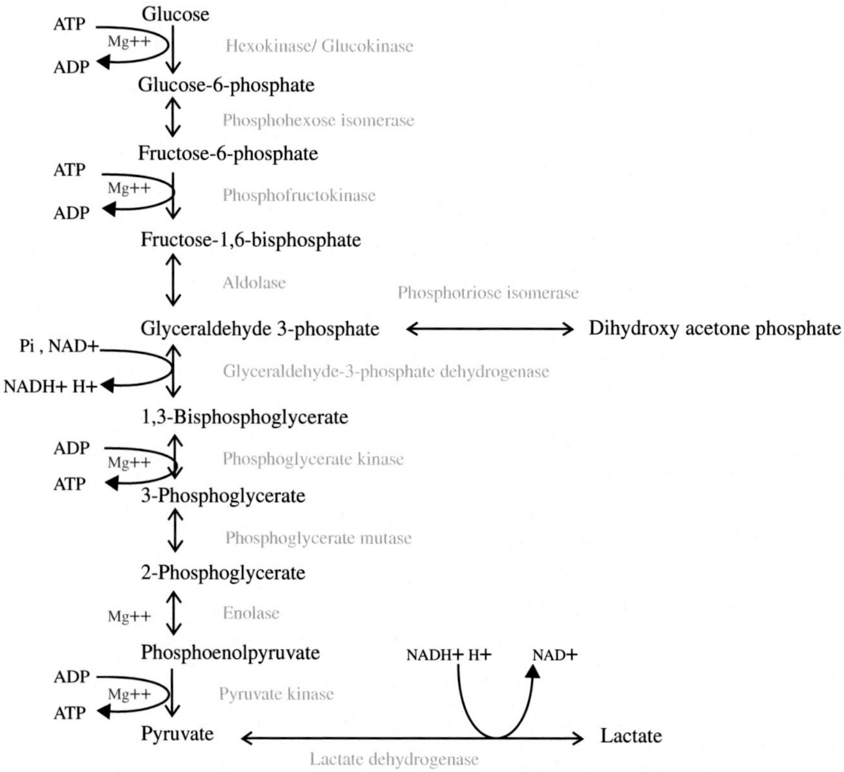

Figure 1.1 Glycolysis pathway.

RHYMING GLYCOLYSIS

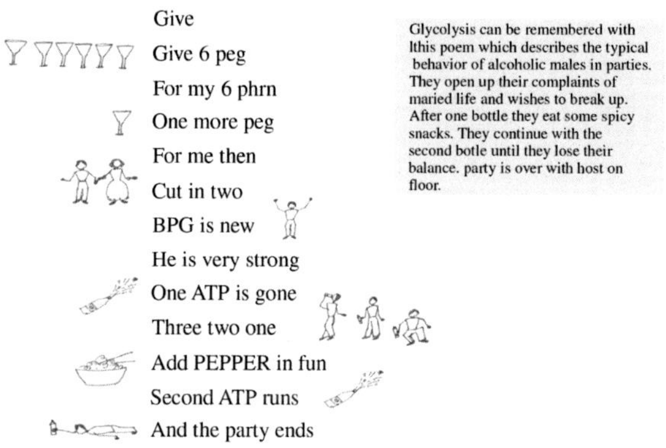
Give
Give 6 peg
For my 6 phrn
One more peg
For me then
Cut in two
BPG is new
He is very strong
One ATP is gone
Three two one
Add PEPPER in fun
Second ATP runs
And the party ends
Glycolysis can be remembered with lthis poem which describes the typical behavior of alcoholic males in parties. They open up their complaints of maried life and wishes to break up. After one bottle they eat some spicy snacks. They continue with the second botle until they lose their balance. party is over with host on floor.

Figure 1.2 A poem to remember the glycolysis intermediates.

What relates to glycolysis intermediates in the poem

Give	Glucose
Give 6 peg	Glucose-6-P
For my 6 phrn (friend)	Fructose-6-P
One more peg	Fructose 1,6-bisP
For me then	
Cut in two	Glyceraldehyde-3-P + DHAP
BPG is new	1,3-Bisphosphoglycerate
He is very strong	
One ATP is gone	3-Phosphoglycerate
Three two one	2-Phosphoglycerate
Add PEPPER in fun	Phosphoenolpyruvate
Second ATP runs	
And the party ends.	Pyruvate

The words in the poem in Fig. 1.2 are related to the intermediates of glycolysis. Red-colored letters are taken from molecules or are short forms of molecules. The "P" in peg in the second and fourth lines indicates the phosphate group, it is cut in two to denote the splitting reaction. "BPG," the name of the protagonist, is a short form of 1,3-bisphosphoglycerate. As it is a high-energy compound, the following line gives the description “He is very strong.” "Three two one" indicates that one phosphate molecule is shifted from the third to the second position. "PEP" in red in the next line is phosphoenolpyruvate, and "P" in the last line is for pyruvate (Fig. 1.3).

GLYCOLYSIS MNEMONIC DIAGRAM

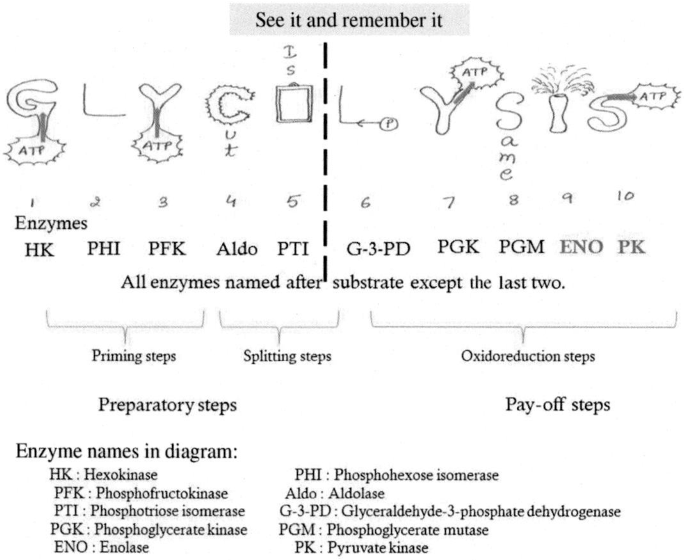

Figure 1.3 Mnemonic diagram to remember the glycolysis reactions.

In glycolysis, glucose (6C) is broken into two pyruvate (3C) molecules.

It has 10 letters in its name and 10 reactions.

Write the word "GLYCOLYSIS" and write the numbers starting from "G" on the left-hand side.

Draw a line at the center of the name dividing it into two parts. The first five are the preparatory steps and the second five are the pay-off or energy-generating steps. Of the preparatory steps, the first three are priming and the subsequent two are splitting steps.

ATP is consumed in steps 1 and 3, so ATP is shown entering at letters "G" and "Y."

The next step is cleavage of fructose 1,6-bisphosphate into glyceraldehyde-3-P and DHAP, indicated by "Cut."

DHAP is isomerized into glyceraldehyde-3-P; this is indicated by the mirror in the "O" of "ISO."

Inorganic phosphate enters along with NAD+ in the following step, therefore "O" is depicted entering "L."

ATP is generated in steps 7 and 10, hence ATP is leaving letters "Y" and "S."

The eighth step is the isomerization reaction in which the molecular formula is "SAME" and only the position of phosphate is changed.

The ninth reaction is dehydration, which is depicted by water coming out of a pipe.

This is how all 10 reactions of glycolysis can be remembered.

CHAPTER 2

Citric Acid Cycle

TRADITIONAL KREB'S CYCLE RECAP

The citric acid cycle utilizes mitochondrial enzymes for final oxidation of carbohydrates, proteins, and fats. Moreover, the Krebs cycle also produces intermediates which are important in gluconeogenesis, lipolysis, neurotransmitter synthesis, etc. Acetyl-CoA from pyruvate of glycolysis, beta oxidation of fatty acids, ketogenic amino acids, and ketones enter this pathway for energy production. The first step is fusion of the acetyl group of acetyl-CoA with oxaloacetate, catalyzed by citrate synthase. CoA-SH and heat are released and citrate is produced. Citrate is isomerized by dehydration and rehydration to isocitrate. The enzyme aconitase catalyzes these two steps using *cis*-aconitate as the intermediate. The next two steps are catalyzed by isocitrate dehydrogenase. Dehydrogenation of isocitrate forms oxalosuccinate, which decarboxylates to alpha-ketoglutarate. Alpha-ketoglutarate is further oxidatively decarboxylated by alpha-ketoglutarate dehydrogenase—a multienzyme complex. Succinyl-CoA is formed in this unidirectional reaction.

Succinate thiokinase converts succinyl-CoA to succinate, while first generating ATP/GTP by substrate-level phosphorylation. Succinate is acted upon by succinate dehydrogenase, requiring FAD and Fe-S proteins to form fumarate. Fumarase adds water to a double bond of fumarate yielding malate. Malate regenerates oxaloacetate by action of NAD+ -dependent malate dehydrogenase, completing the cycle.

Coenzymes including FAD and NAD+ are reduced in the Krebs cycle, which transfers electrons by the electron transport chain with oxygen as the final acceptor. Three NADH+ and one $FADH_2$ are generated in one cycle which, on entering the electron transport chain, yields 10 ATP. These include one ATP produced by succinate

Sweet Biochemistry: Remembering Structures, Cycles, and Pathways by Mnemonics
DOI: http://dx.doi.org/10.1016/B978-0-12-814453-4.00002-9

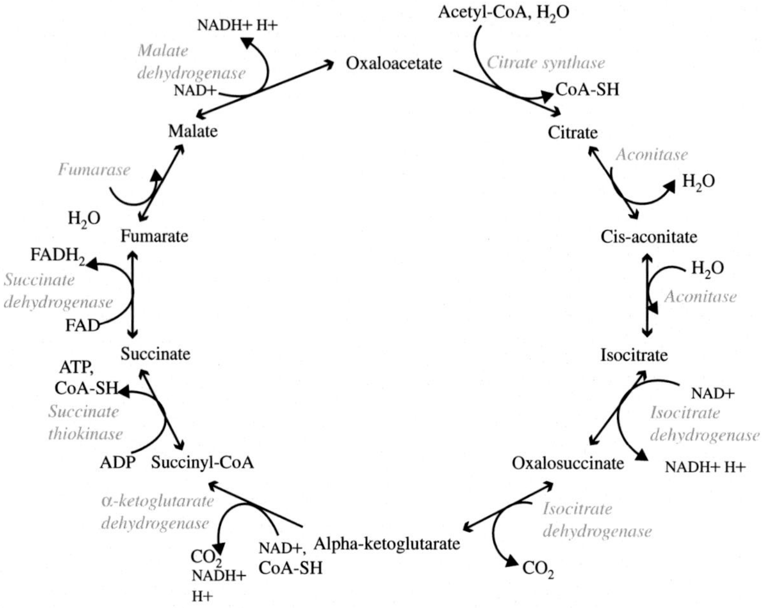

Figure 2.1 The citric acid cycle.

thiokinase at the substrate level. Two carbon atoms are lost in this cycle by decarboxylation, although these are not the same atoms entering as acetyl-CoA. Vitamins such as riboflavin, niacin, and thiamine work as coenzymes in this cycle, while pantothenic acid forms the CoA part of acetyl-CoA.

Fluoroacetate inhibits aconitase, arsenite inhibits alpha-ketoglutarate, and malonate inhibits succinate dehydrogenase.

The Krebs cycle is one of the key pathways responsible for energy production, with the liver being the principal site. In the case of enzyme defects in the Krebs cycle, ATP production is hampered to a great extent, leading to severe brain damage (Figs. 2.1–2.3).

STORY MNEMONIC OF CITRIC ACID CYCLE

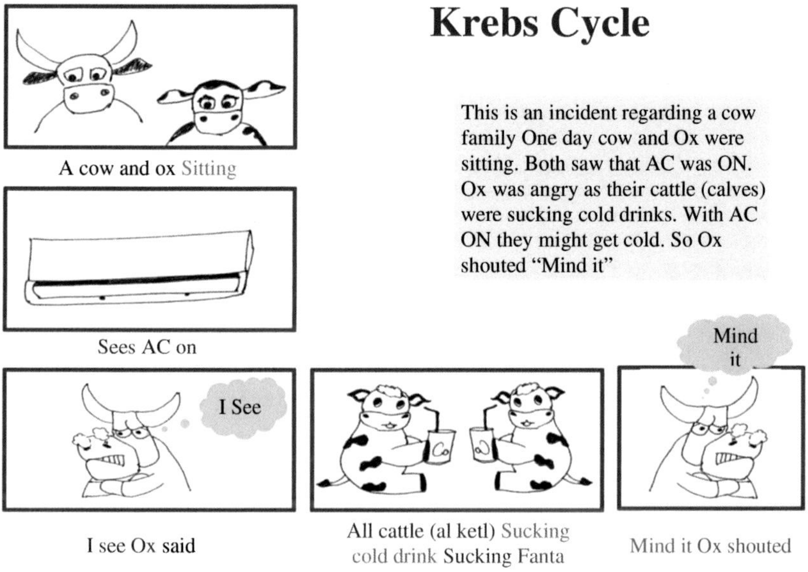
Krebs Cycle
This is an incident regarding a cow family One day cow and Ox were sitting. Both saw that AC was ON. Ox was angry as their cattle (calves) were sucking cold drinks. With AC ON they might get cold. So Ox shouted "Mind it"
A cow and ox Sitting
Sees AC on
I See
I see Ox said
All cattle (al ketl) Sucking cold drink Sucking Fanta
Mind it
Mind it Ox shouted

Figure 2.2 The mnemonic story of the citric acid cycle.

What relates to Krebs cycle intermediates in the story

		Enzymes acting on	
A cow and ox	Acetyl CoA+ oxaloacetate	CS	1
1. Sitting (citrate)	Citrate	Aco	2
2. Sees AC on	Cis-aconitate		3
3. I see	Isocitrate	ICD	4
4. Ox said	Oxalosuccinate		5
5. All cattle (al ketl)	Alpha-ketoglutarate	KGD	6
6. Sucking cold drink	Succinyl CoA	ST	7
7. Sucking	Succinate	SD	8
8. Fanta	Fumarate	F	9
9. Mind it	Malate	MD	10
10. Ox shouted	Oxaloacetate		

The first two enzymes are names related to products and the remaining eight to substrate. (In glycolysis the last two enzymes are related to products)

PICTURE MNEMONIC OF CITRIC ACID CYCLE

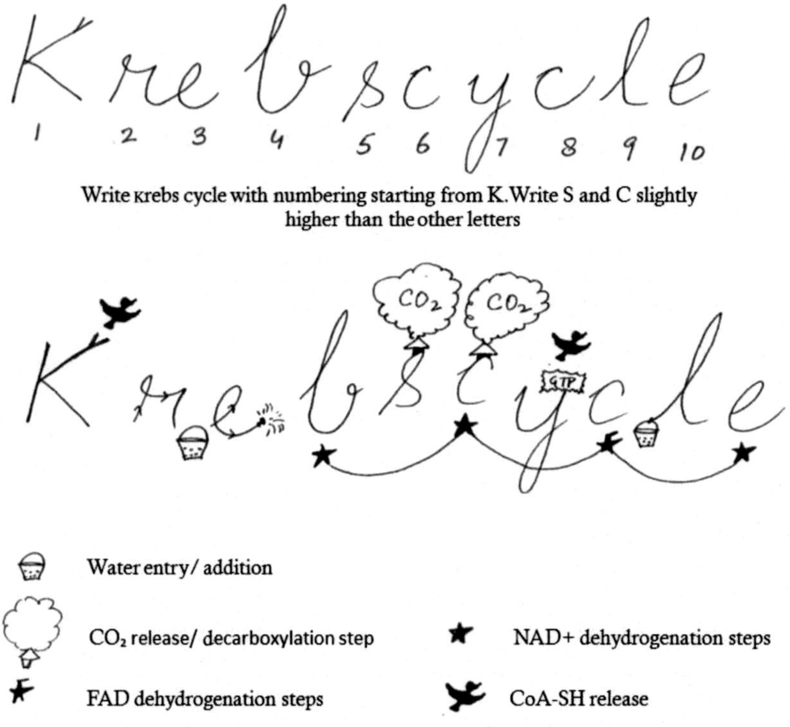

Figure 2.3 A second way of remembering the Krebs cycle, with a picture mnemonic for the pathway reactions.

LEARNING THE CITRIC ACID CYCLE MNEMONIC

In the Krebs cycle, there are 10 letters and 10 reactions. Write "Krebs cycle" and start numbering from "K." Write "S" and the first "C" a little higher than the rest of the letters.

A crow (CoAA in Hindi) is flying from letters "K" and "Y," indicating CoA-SH release.

Dehydration and rehydration take place sequentially, so that water is entering at "R" and exiting from "E."

You can see a ★ below "B" which indicates the NAD+ -dependent dehydrogenation step.

The next two letters, "S" and "C," are flying a little higher because of the CO_2 release in decarboxylation. "C" also has a ★

dehydrogenation reaction denoted by "Y" carrying "GTP" on its shoulder, which is produced in substrate-level phosphorylation.
The second "C" is again a dehydrogenation reaction with a different star. This is because it is FAD-dependent dehydrogenation.★

"L" is attached to a bucket with water for hydration.

The final dehydrogenation (NAD+ -dependent ★) hangs from "E."

Therefore, in the Krebs cycle there are four dehydrogenation reactions, which are alternatively placed; the third one being FAD-dependent.

CHAPTER 3

Electron Transport Chain

TRADITIONAL ELECTRON TRANSPORT CHAIN RECAP

The electron transport chain is a mitochondrial pathway in which electrons move across a redox span of 1.1 V from NAD+/NADH to O_2/H_2O. Three complexes are involved in this chain, namely, complex I, complex III, and complex IV. Some compounds like succinate, which have more positive redox potential than NAD+/NADH, can transfer electrons via a different complex—complex II.

Coenzyme Q or simply Q can travel within membrane, while Cyt C is a soluble protein. Flavoproteins are components of complexes I and II and Fe-S is present in complexes I, II, and III. The Fe atom present in Fe-S complexes helps in electron transfer by shifting from Fe^{2+} to Fe^{3+} states.

Electrons are transferred from NADH to FMN, from where they enter Fe-S complexes. From Fe-S they move to Q, which carries them to complex III. In complex III electrons are received by cytochrome c1 and cytochrome b and sent to cyt c.

Cyt c transfers electrons to complex IV, where they are passed from the copper center to heme a, heme a3, and a second copper center. Finally, from complex IV, electrons are received by molecular oxygen and water is formed.

Complex II is used when electrons enter via FAD and then go to Fe-S centers to Q. Succinate conversion to fumarate is a common source for these electrons.

It is important to note that Q can carry two electrons, while cyt c can transfer only one, hence when one QH_2 is oxidized, two molecules of cyt c are reduced.

During the flow of electrons across these complexes, protons are sent to intermitochondrial membrane space where a proton gradient is

Sweet Biochemistry: Remembering Structures, Cycles, and Pathways by Mnemonics
DOI: http://dx.doi.org/10.1016/B978-0-12-814453-4.00003-0

generated. Protons in the process of running through available spaces rotate ATP synthase enzymes embedded in mitochondrial membrane and ADP is phosphorylated. Thus, the oxidation of reducing equivalents is coupled to phosphorylation of ADP. This is termed oxidative phosphorylation. Molecules which interfere with ETC or coupling of oxidative phosphorylation are often fatal, for example, cyanide and carbon monoxide. Genetic disorders involving ETC components lead to decreased ATP production and present with myopathy, fatigue, and lactic acidosis (Figs. 3.1 and 3.2).

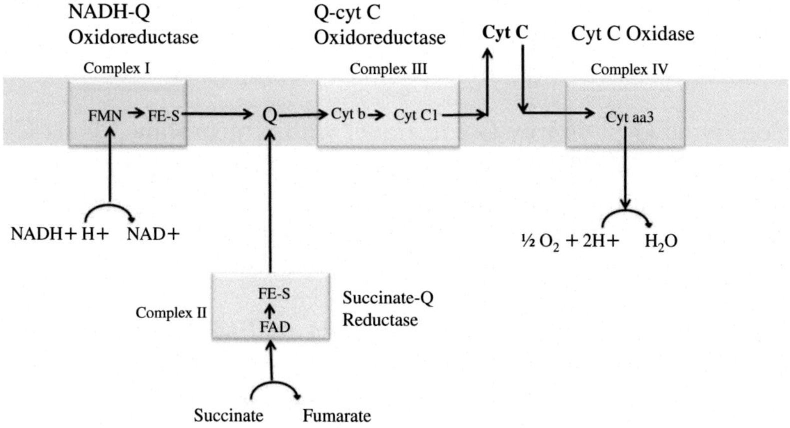

Figure 3.1 Electron transport chain.

STORY MNEMONIC OF ETC

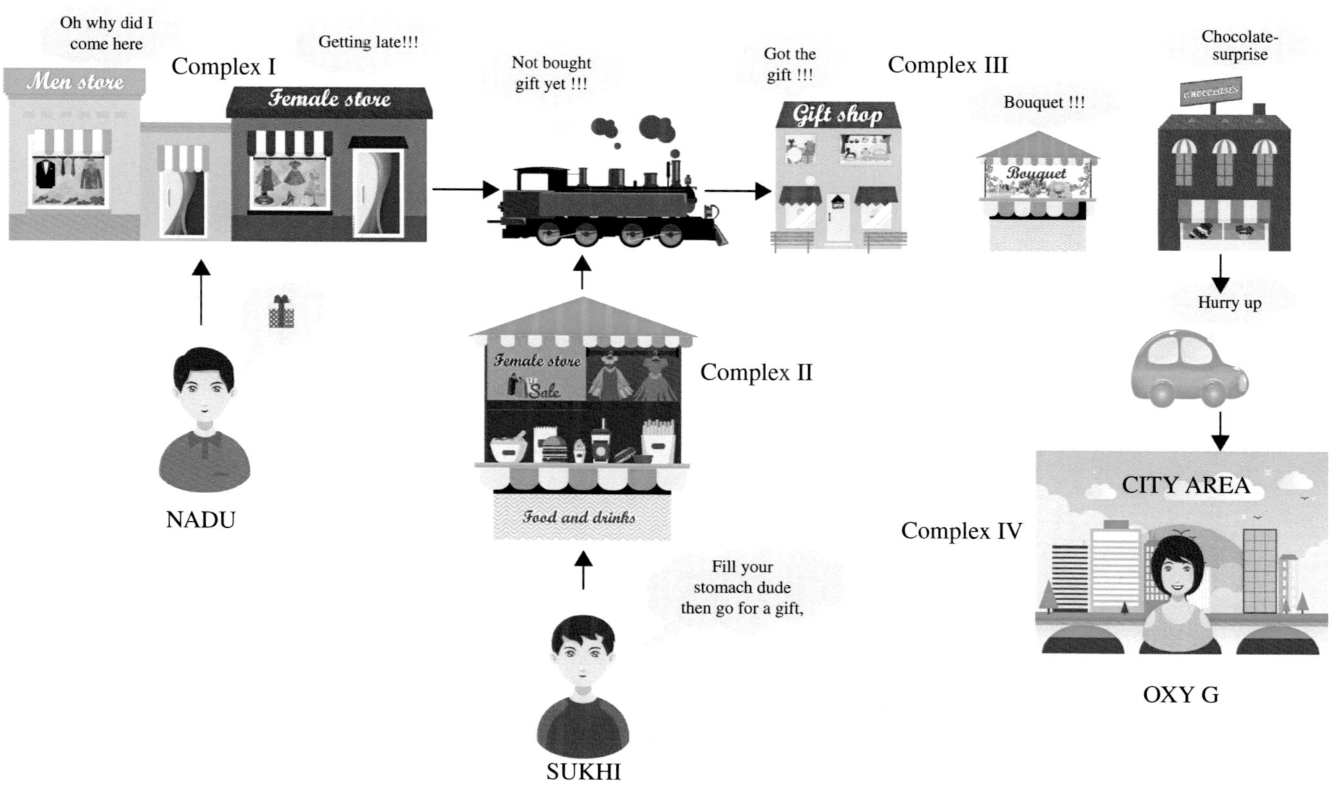

Figure 3.2 Electron transport chain mnemonic story.

The Story of ETC

NADU wants to give a gift to Oxy G, his girlfriend. As a reflex he enters a men's store. He then realizes his mistake and proceeds to a female store. He has not yet purchased the gift and the time has come for him to board his train. He boards the train towards his destination. He again searches for a gift in a female store and buys the gift. He adds a bouquet and chocolate from city shops. Then NADU takes a car and reaches city area 3, where Oxy G is waiting. Sukhi (NADU's friend) had suggested to him that they have food and drinks and then go to the women's store to buy gifts.

WHAT RELATES TO THE ELECTRON TRANSPORT CHAIN IN THIS STORY

NADU → NADH
For MeN store → FMN
Female Store → Fe-S
Train → Coenzyme Q
City bouquet shop → Cyt b
City chocholate shop → Cyt c1
Car → Cyt C
City AreA3 → Cyt a a3
Oxy G → oxygen
Sukhi → succinate
Food And Drinks → FAD

CHAPTER 4

Beta Oxidation of Fatty Acids

TRADITIONAL BETA OXIDATION OF FATTY ACIDS RECAP

Fatty acid oxidation is the mitochondrial aerobic process of breaking down fatty acids into acetyl-CoA units. Fatty acids move in this pathway as CoA derivatives utilizing NAD and FAD.

Fatty acids are activated before oxidation utilizing ATP in the presence of CoA-SH and acyl-CoA synthetase. Long-chain acyl-CoA enters mitochondria bound to carnitine. Inside mitochondria beta oxidation of fatty acids takes place in which two carbon atoms are removed in the form of acetyl-CoA from acyl-CoA at the carboxyl terminal. The bond is broken between the second carbon/beta carbon and the third carbon/gamma carbon, hence the name beta oxidation.

FAD-dependent dehydrogenation is started by acyl-CoA dehydrogenase, which results in a double-bond formation between C2 and C3. $FADH_2$ is generated in this reaction. In the next step water is added by enoyl-CoA hydratase on a double bond forming 3-hydroxyacyl-CoA. Secondly, NAD+-associated dehydrogenation by 3-hydroxyacyl-CoA dehydrogenase converts the hydroxy group on C3 to a keto group yielding 3-ketoacyl-CoA.

Thiolase cleaves the bond between C2 and C3, releasing acetyl-CoA and acyl-CoA which is two carbon atoms shorter than the starting molecule. This new acyl-CoA again enters the same pathway.

In this way acetyl-CoA is sequentially removed from acyl-CoA until two acetyl-CoA molecules are left. Odd-chain fatty acids leave acetyl-CoA and propionic acid on completion. Acetyl-CoA is directed to the citric acid cycle for further oxidation. One cycle of beta oxidation releases one $FADH_2$ molecule and one NADH+ molecule, which causes the synthesis of four high-energy phosphate bonds in electron transport chain (ETC). Beta oxidation of fatty acids thus provides a large number of ATP (one palmitic acid molecule provides approximately 106 mol ATP). Therefore, this process is stimulated in

Sweet Biochemistry: Remembering Structures, Cycles, and Pathways by Mnemonics
DOI: http://dx.doi.org/10.1016/B978-0-12-814453-4.00004-2

starvation, when glucose level falls. In patients with diabetes mellitus, when glucose is not able to enter cells, beta oxidation of fatty acid is the rescuer. Released acetyl-CoA enters ketone bodies, acting as an energy substrate for many tissues, including the brain. By providing substrates for gluconeogenesis, beta oxidation of fatty acids prevents hypoglycemia. This becomes evident in defects of the beta oxidation pathway, for example in carnitine deficiency, CPT enzyme defects, or hypoglycin poisoning (Figs. 4.1–4.3).

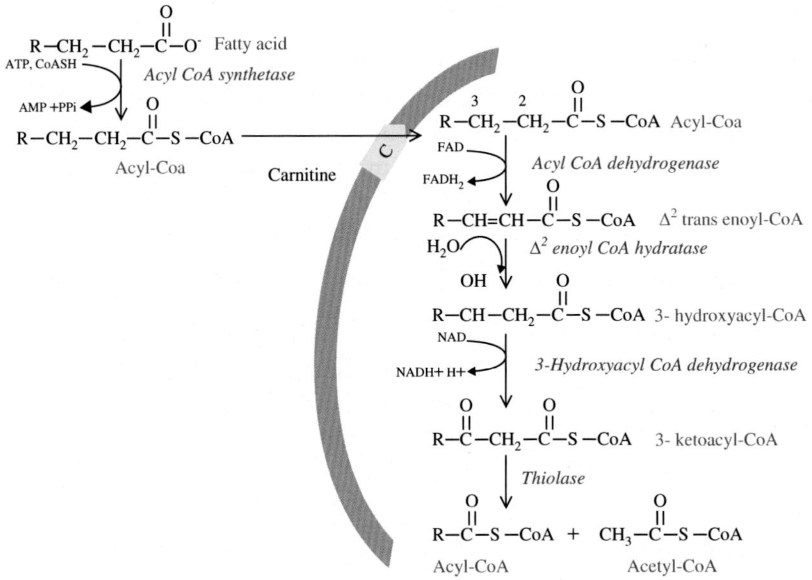

Figure 4.1 Beta oxidation of fatty acids.

PICTURE MNEMONIC OF BETA OXIDATION OF FATTY ACIDS

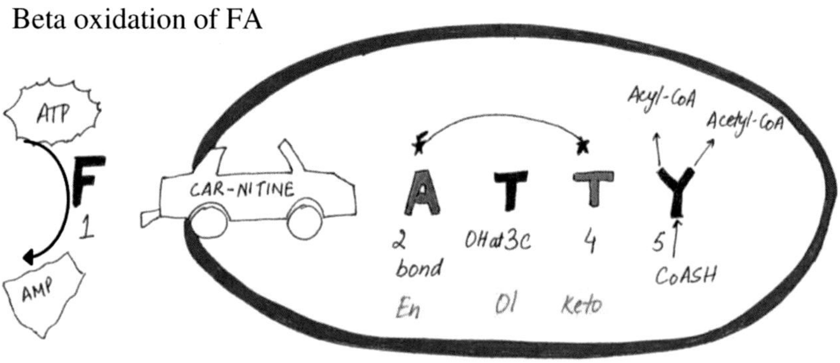

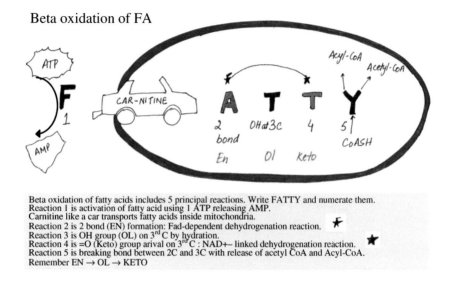

Beta oxidation of fatty acids includes 5 principal reactions. Write FATTY and numerate them.
Reaction 1 is activation of fatty acid using 1 ATP releasing AMP.
Carnitine like a car transports fatty acids inside mitochondria.
Reaction 2 is 2 bond (EN) formation: Fad-dependent dehydrogenation reaction. ★
Reaction 3 is OH group (OL) on 3rd C by hydration.
Reaction 4 is =O (Keto) group arival on 3rd C : NAD+– linked dehydrogenation reaction. ★
Reaction 5 is breaking bond between 2C and 3C with release of acetyl CoA and Acyl-CoA.
Remember EN → OL → KETO

Figure 4.2 Illustrated mnemonic for beta oxidation of fatty acids. Important reactions are summarized in this figure.

SIMPLE WAY TO LEARN BETA OXIDATION OF FATTY ACIDS

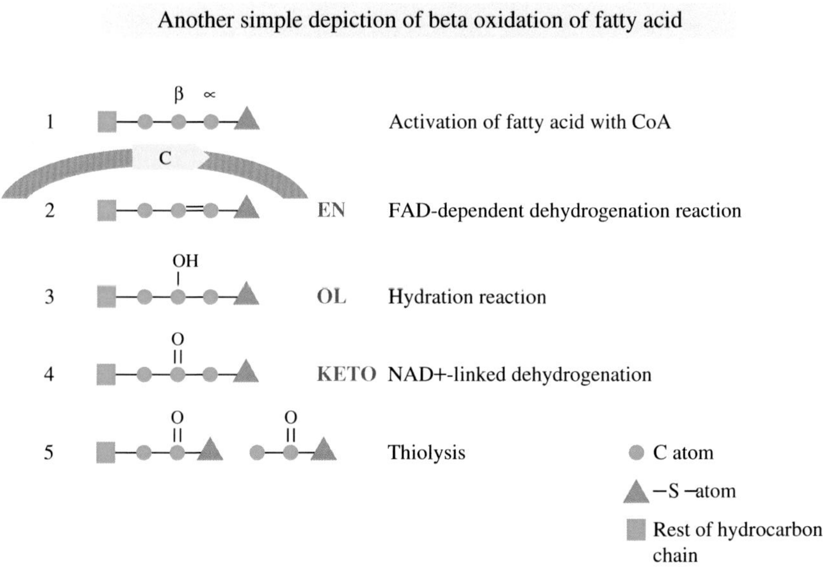

Figure 4.3 Another simple way to understand the basics of the reactions of beta oxidation. The yellow block with "C" inside represents carnitine, which transports long-chain fatty acids.

CHAPTER 5

Fatty Acid Biosynthesis

TRADITIONAL FATTY ACID BIOSYNTHESIS RECAP

Fatty acids are synthesized in cytoplasm with the participation of ATP, NADPH, biotin, HCO_3^-, and Mn^{++}. The main sites for this pathway are the liver, kidneys, brain, lungs, mammary glands, and adipose tissue, where the HMP pathway can provide an ample amount of NADPH. Acetyl-CoA is the two-carbon building block molecule which is added to malonyl-CoA (formed by carboxylation of acetyl-CoA). The enzyme catalyzing this reaction is the regulatory enzyme of fatty acid synthesis—acetyl-CoA carboxylase. This process is cytoplasmic, while beta oxidation takes place in the mitochondria. The main substrate for biosynthesis is glucose, which is why a high-carbohydrate diet can promote lipogenesis. Fatty acid synthesis is a very important pathway in view of the alarming increase in obesity and diabetes mellitus. Fatty acids are precursors of eicosanoids, complex lipids, membrane lipids, and secondary messengers of signal transduction regulating cell functions.

The first step is transfer of acetyl-CoA to the cysteine-SH group by acetyl transacylase and malonyl-CoA on 4'-phosphopantetheine of ACP by malonyl transacylase. These two sites belong to different units of the dimer enzyme complex called fatty acid synthase. This complex enzyme has seven enzyme activities on monomers, and hands over substrate from one component enzyme to the other. This improves the efficiency of the reaction.

In the second step, the acetyl group joins the malonyl residue by 3-ketoacyl synthase and CO_2 is released, yielding 3-ketoacyl. One site then becomes free. 3-Ketoacyl is reduced in third step utilizing NADPH and 3-ketoacyl reductase, forming 3-hydroxyacyl. The fourth step is dehydration catalyzed by hydratase. This step creates a double bond between C2 and C3. This bond is reduced by enoyl reductase enzyme. The second reduction also needs NADPH.

Sweet Biochemistry: Remembering Structures, Cycles, and Pathways by Mnemonics
DOI: http://dx.doi.org/10.1016/B978-0-12-814453-4.00005-4

Finally acyl-S enzyme is formed. This acyl residue is then shifted to the empty cysteine-SH group and a new malonyl residue arrives at –SH of 4'-phosphopantetheine.

The above five reactions are repeated until the desired length is achieved. Thioesterase releases the completed fatty acid by hydration.

Linoleic and linolenic fatty acids are called essential fatty acids because they cannot be synthesized by humans (Figs. 5.1 and 5.2).

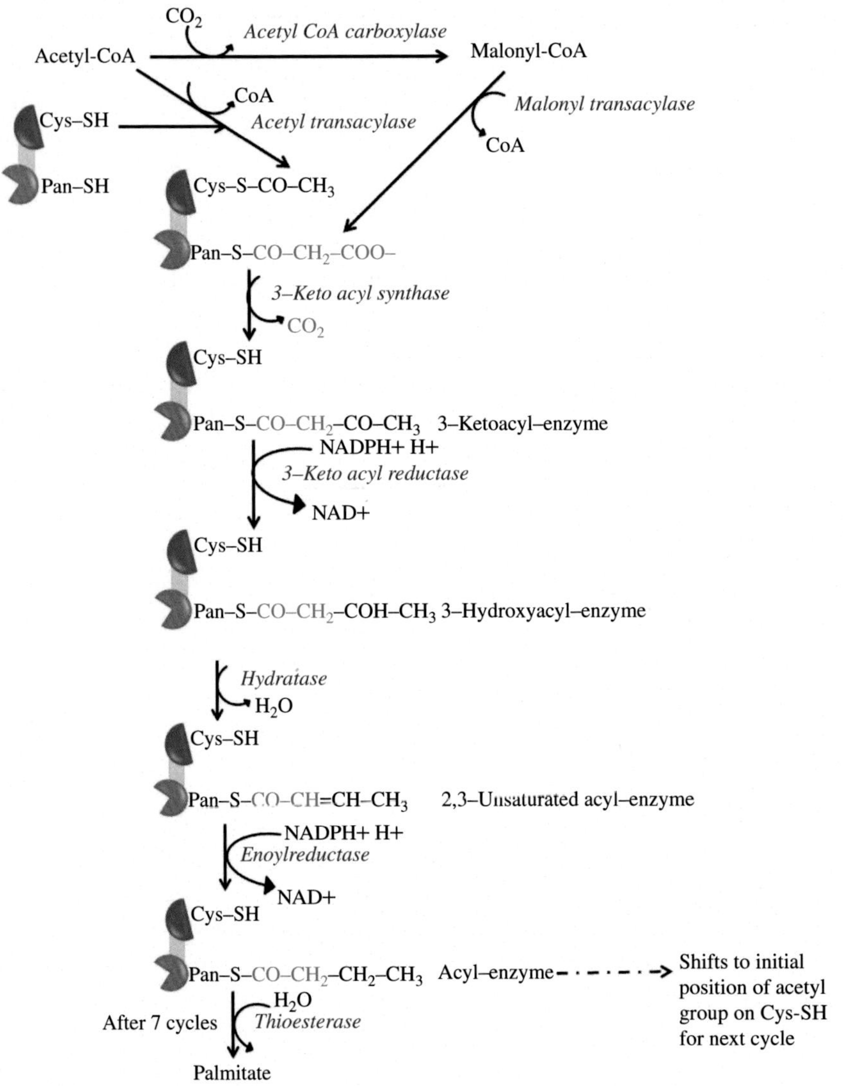

Figure 5.1 Fatty acid biosynthesis.

STORY OF FATTY ACID SYNTHESIS

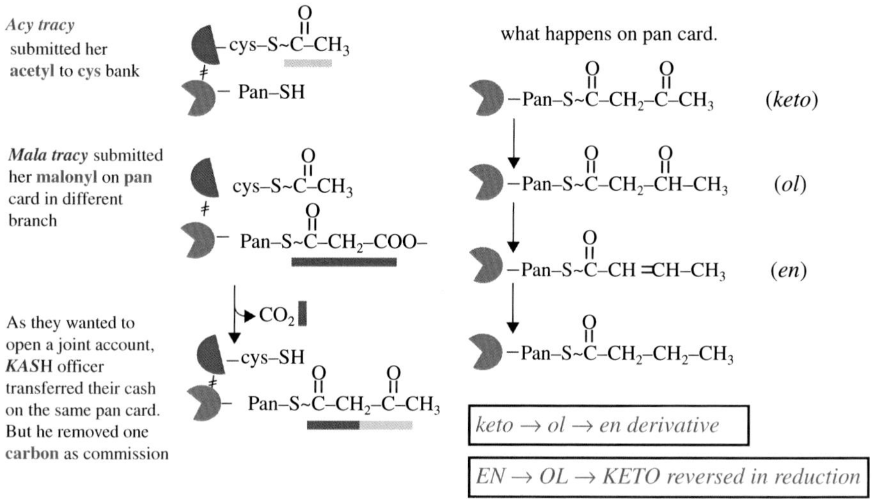

Figure 5.2 The fatty acid synthesis story.

Fatty acid synthesis can be compared with depositing money (carbon atoms) into a bank. When the desired amount has been gathered the money can be taken out. Two enzyme sisters, Acy Tracy and Mala Tracy, deposit their cash (carbon atoms) into a bank. Acy Tracy deposits her cash in the CYS bank. Mala Tracy deposits her cash on the same pan card in a different branch. Then, both wanted to shift their money into a joint account, so the Cash (KAS) officer transferred Acy's carbon onto the same pan card but kept one carbon as commission. On the pan card the cash followed the sequence of keto → ol → en (Fig. 5.3).

What relates to fatty acid biosynthesis in story?

	Ketoacyl synthase
	Acyl carrier protein (ACP)
Acy tracy	Acetyl transacylase
cys bank	Cysteine —SH group of Ketoacyl synthase
Mala tracy	Malonyl transacylase
pan card	—SH on the 4′-phosphopantetheine of ACP
KASH	3-ketoacyl synthase

RELATION BETWEEN FATTY ACID SYNTHESIS AND OXIDATION

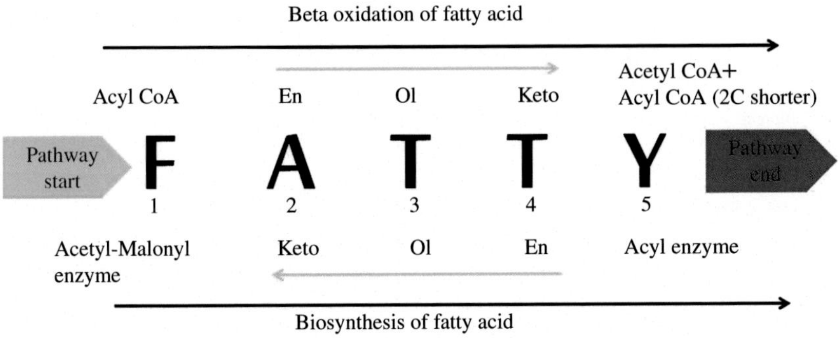

Figure 5.3 Beta oxidation and synthesis of fatty acids.

Fatty acid synthesis and oxidation are not exactly opposite processes, and their locations are also different. Synthesis is cytosolic, and oxidation is mitochondrial. However, it has been noted that fatty acid synthesis and beta oxidation both have five important steps. The two processes run in opposite directions as shown above.

In beta oxidation fatty acid is activated by CoASH and then en → ol → keto derivatives are formed. Acetyl CoA is then removed.

The opposite happens in synthesis, where acetyl-CoA is added to the growing acyl group and keto → ol → en derivatives are formed and finally reduced.

CHAPTER 6

Cholesterol Structure

TRADITIONAL CHOLESTEROL STRUCTURE RECAP

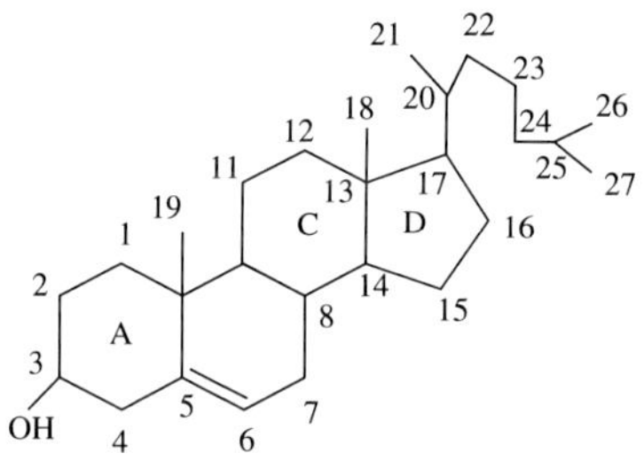

Cholesterol is the parent molecule of steroids. Cholesterol is a 27-carbon compound characterized by a steroid ring. This steroid nucleus includes three cyclohexane rings fused in a phenanthrene arrangement along with a cyclopentane. A double bond is present between the 5th and 6th carbon atoms. One hydroxyl group is present at the 3rd carbon and it participates in hydrogen bond formation with carbonyl oxygen of the phospholipid head. The steroid ring and alcohol(ol) group contribute to the name sterol.

Figure 6.1 Cholesterol structure.

Cholesterol is a 27-carbon compound characterized by a steroid ring. This steroid nucleus includes three cyclohexane rings fused in a phenanthrene arrangement, along with a cyclopentane. A double bond is present between the 5th and 6th carbon atoms (Fig. 6.1).

One hydroxyl group is present at the 3rd carbon and it participates in hydrogen bond formation with carbonyl oxygen of the phospholipid head. The steroid ring and alcohol(ol) group contribute to the name sterol. Due to the presence of asymmetric carbons in the nucleus, chair, and boat isomers are possible; the chair form is more stable and common.

Due to its unique bulky nucleus, cholesterol is abundantly present in plasma membrane. The hydroxyl group at carbon 3 forms a hydrogen bond with the carbonyl oxygen atom of the phospholipid head group, whereas the hydrophobic tail adjusts to the lipophilic core of the membrane. Cholesterol inserts itself perpendicularly to the membrane plane of the phospholipid bilayer, where it interferes in reactions between fatty acids. The formation of lipid rafts is another important biological property of cholesterol. Lipid rafts are stiff complexes of cholesterol with phospholipids and proteins, they make membrane

Sweet Biochemistry: Remembering Structures, Cycles, and Pathways by Mnemonics
DOI: http://dx.doi.org/10.1016/B978-0-12-814453-4.00006-6

around them less fluid but some proteins present on lipid rafts act as receptors and channels.

Cholesterol is the parent molecule of steroids like glucocorticoids, mineralocorticoids, reproductive hormones, and vitamin D. Bile acids are catabolic products of cholesterol, which play a vital role in fat emulsification, which is required for fat and fat-soluble vitamin absorption.

HONEYCOMB HOUSE- CHOLESTEROL

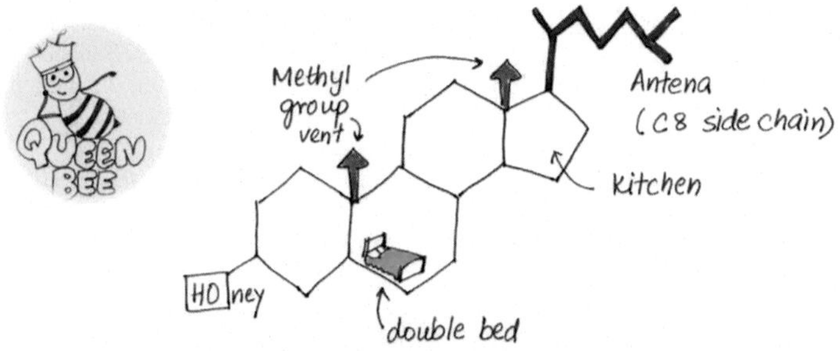

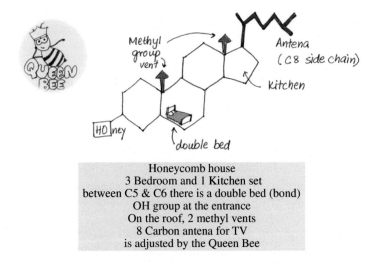

Honeycomb house
3 Bedroom and 1 Kitchen set
between C5 & C6 there is a double bed (bond)
OH group at the entrance
On the roof, 2 methyl vents
8 Carbon antena for TV
is adjusted by the Queen Bee

Figure 6.2 Cholesterol structure mnemonic with a rhyme sung by the Queen Bee.

WHAT RELATES TO CHOLESTEROL IN THE HONEYCOMB HOUSE?

3 bedrooms → 3 cyclohexane rings
1 kitchen → one cyclopentane ring
Double bed → double bond between C5 and C6
OH group at entrance → OH group at C3
On the roof, 2 methyl vents → methyl groups at C10 and C13
Antena → 8-carbon side chain on C17

Fig. 6.2 greatly helps in remembering the cholesterol structure if you visit each corner of the house vigilantly—Try singing the rhyme.

CHAPTER 7

Cholesterol Synthesis

TRADITIONAL CHOLESTEROL SYNTHESIS RECAP

Cholesterol synthesis is an expensive process for cells in terms of energy. This pathway takes place in cytoplasm. The liver and intestines are major contributors to endogenous production. Acetyl-CoA units are joined to form a 30-carbon compound and then three carbons are removed to produce cholesterol which has 27 carbon atoms.

The cholesterol synthesis steps can be divided into:

1. Mevalonate synthesis
2. Isopentenyl phosphate synthesis
3. Squalene formation
4. Lanosterol synthesis
5. Cholesterol formation.

Two acetyl-CoAs combine to form acetoacetyl-CoA, releasing CoA-SH in the presence of thiolase. Acetyl-CoA also condenses to form 3-hydroxy-3-methylglutaryl-CoA (HMG-CoA) catalyzed by HMG-CoA synthase. These enzymes are different from the enzymes used for ketone body synthesis in mitochondria. HMG-CoA is reduced by HMG-CoA reductase using NADPH to mevalonate. This enzyme is the regulatory enzyme of the pathway, and is inhibited by statins—the best lipid-lowering drugs.

Mevalonate is phosphorylated by three kinases sequentially utilizing three ATPs and is then decarboxylated to form isopentenyl diphosphate.

Isopentenyl diphosphate (5C) isomerizes to 3,3-dimethylallyl diphosphate (5C) by shifting a double bond and then condensation with isopentenyl diphosphate forms geranyl diphosphate (10C). Another isopentenyl diphosphate molecule joins to form the 15C compound, farnesyl diphosphate. Two such 15C molecules fuse to form 30C squalene.

Sweet Biochemistry: Remembering Structures, Cycles, and Pathways by Mnemonics
DOI: http://dx.doi.org/10.1016/B978-0-12-814453-4.00007-8

Squalene is oxidized to squalene 2,3-epoxide by squalene epoxidase. During cyclization to lanosterol, a methyl group shifts from C14 to C13 and from C8 to C14.

The methyl groups on C14 and C4 are removed to form 14-desmethyl lanosterol and then zymosterol. The double bond at C8–C9 is subsequently shifted to C5–C6 in two steps, forming desmosterol. The final step is the reduction of the double bond of the side chain yielding a cholesterol molecule (Fig. 7.1).

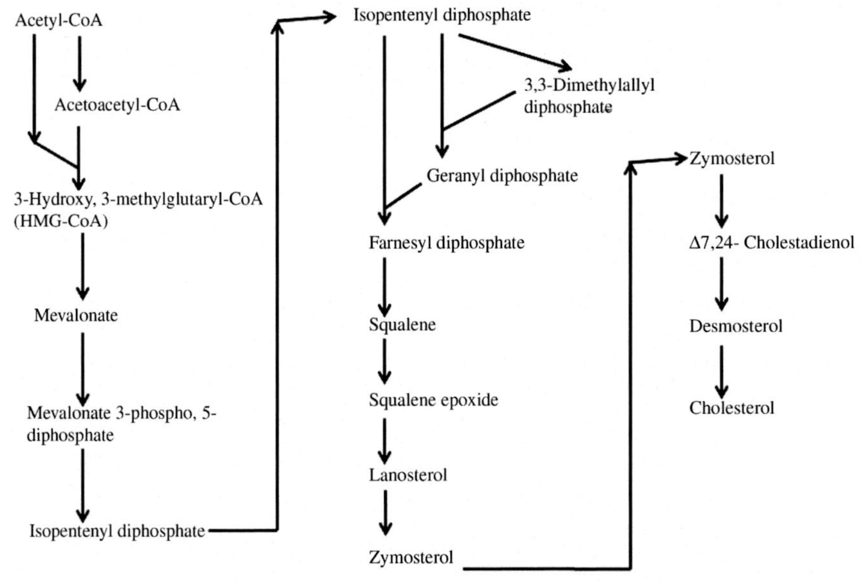

Figure 7.1 The basics of cholesterol synthesis.

CHOLESTEROL SYNTHESIS STEPS

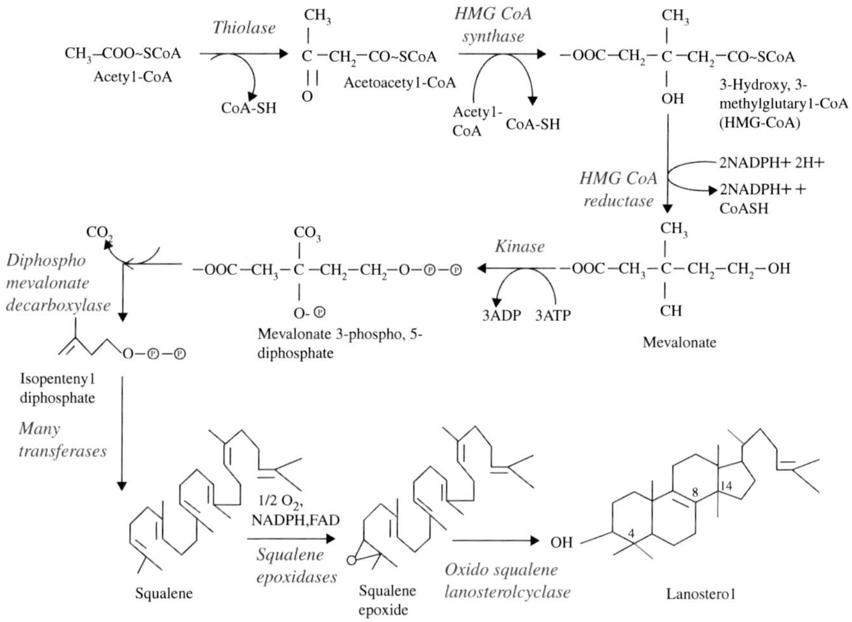

Figure 7.2 Cholesterol synthesis steps, Part 1.

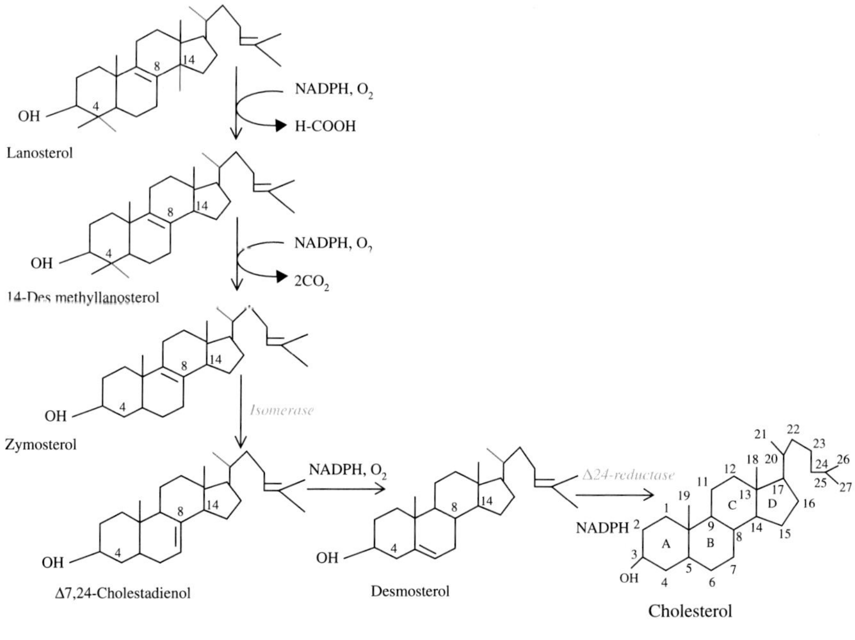

Figure 7.3 Cholesterol synthesis steps, Part 2.

Two acetyl-CoA units continue to join until they form a 30-carbon structure, squalene (Fig. 7.2). Squalene is oxidized, cyclized, decarboxylated, and again reduced to yield cholesterol (Figs. 7.3 and 7.4).

STORY OF CHOLESTEROL SYNTHESIS

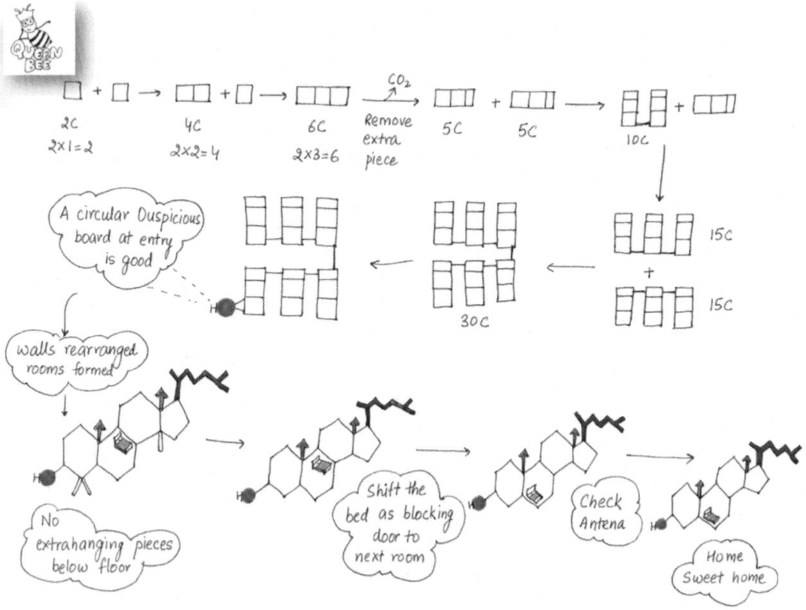

Figure 7.4 The Queen Bees narrates the story of making her honeycomb house.

QUEEN BEE'S HONEYCOMB HOUSE

The Queen Bee decides to make her honeycomb house with three bedrooms and one kitchen. She started joining bricks of acetyl-CoA (2C) in the following manner:

$$2C + 2C = 4C$$
$$4C + 2C = 6C$$
$$6C - 1C = 5C$$
$$5C + 5C = 10C$$
$$10C + 5C = 15C$$
$$15C + 15C = 30C$$

Once she reaches 30C, the material for her house is complete. She starts by hanging an Ouspicious (auspicious) board at the entrance to the house. Then she rearranges the walls in such a way that three hexagonal rooms and one pentagonal kitchen are formed. She discards the extra hangings below the floor for a smooth look. Next, she moves her bed, which was obstructing the door to the third room. Finally, when the house is the ready, she corrects her TV antenna's settings.

CHOLESTEROL SYNTHESIS (TABLE 7.1)

Table 7.1 Correlation Between the Queen Bee's story and Cholesterol Synthesis

What the Queen B does	What this correlates to in the cholesterol pathway
2C	Acetyl-CoA
4C	Acetoacetyl-CoA
6C	HMG-CoA
5C	Isoprenoid unit
10C	Geranyl
15C	Fernesyl
30C	Squalene
Ouspicious (auspicious) board at the entrance of house	Squalene is oxidized to squalene 2,3-epoxide
She rearranges the walls	Cyclization
She discards the extra hangings below the floor	The methyl groups on C14 and C4 are removed to form 14-desmethyl lanosterol
Next she moved her bed	The double bond at C8–C9 is subsequently shifted to C5–C6
She corrected her TV antenna	Reduction of the double bond of the side chain

CHAPTER 8

Heme Synthesis

TRADITIONAL HEME SYNTHESIS RECAP

Hemoproteins play a vital role in cellular functions from gaseous exchange to redox reductions. Biologically significant hemoproteins include hemoglobin, myoglobin, catalase, cytochrome c, and cytochrome p450. Heme synthesis starts in mitochondria with the condensation of succinyl-CoA with the amino acid glycine, activated by pyridoxal phosphate. ALA synthase catalyzes this irreversible reaction forming an intermediate amino-ketoadipic acid. ALA synthase is the rate-limiting enzyme of heme synthesis. Two forms of ALA synthase are found: erythroid (ALAS2) and hepatic (ALAS1). ALA molecules enter the cytoplasm where their union in the presence of ALA dehydratase yields porphobilinogen (PBG) and water molecules. ALAD is inhibited by lead, and heme synthesis is inhibited leading to anemia. Four PBG molecules are joined by uroporphyrinogen I synthase (PBG deaminase) as a linear tetrapyrrole called hydroxymethylbilane (HMB). Linear tetrapyrrole cyclizes to form a ring known as uroporphyrinogen III (UPG) with participation of uroporphyrinogen III synthase. Uroporphyrinogen III has one asymmetric side chain.

All acetyl groups of UPG are coverted to methyl groups by decarboxylation and coproporphyrinogen III (CPG) is generated. CPG is acted upon in mitochondria by CPG oxidase, which decarboxylates and oxidizes two propionic side chains to vinyl groups. Protoporphyrinogen thus formed is further oxidized to protoporphyrins. Molecular oxygen is required for conversion of CPG to protoporphyrins. Finally, iron is incorporated to generate heme.

Sweet Biochemistry: Remembering Structures, Cycles, and Pathways by Mnemonics
DOI: http://dx.doi.org/10.1016/B978-0-12-814453-4.00008-X

The heme synthesis pathway is carried out by bone marrow (major contribution) and the liver. Heme, the product of this pathway, regulates its synthesis by decreasing synthesis of ALAS1 by negative aporepressor feedback.

The porphyrias are a group of genetic disorders characterized by deficiency of enzymes of heme synthesis. Heme synthesis is affected in ALAS and ALAD deficiency (Figs. 8.1–8.3).

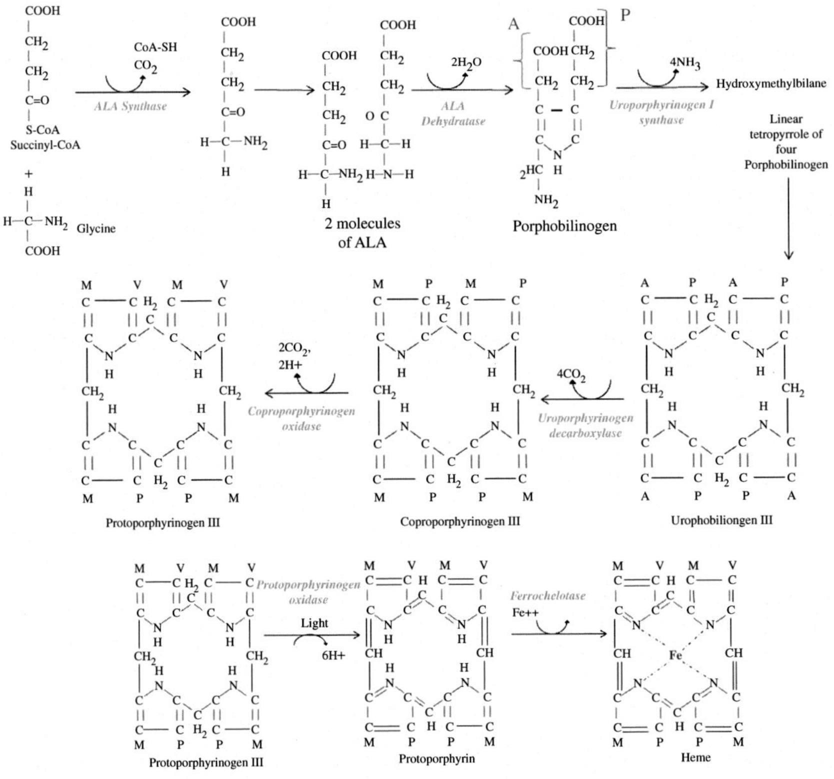

Figure 8.1 Heme synthesis steps.

HE-ME DANCE

Imagine you are a dance instructor preparing a group dance of HE-ME dance. Males are suCoA and females are glycine. You complete the dance in 8 instructions to dancers.

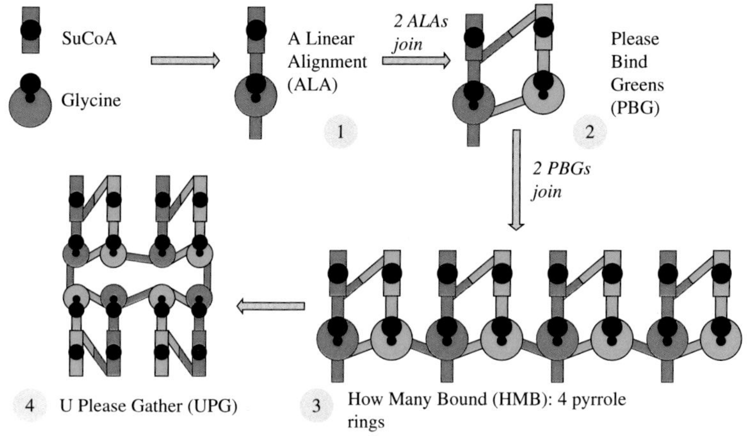

Figure 8.2 Heme dance, Part 1.

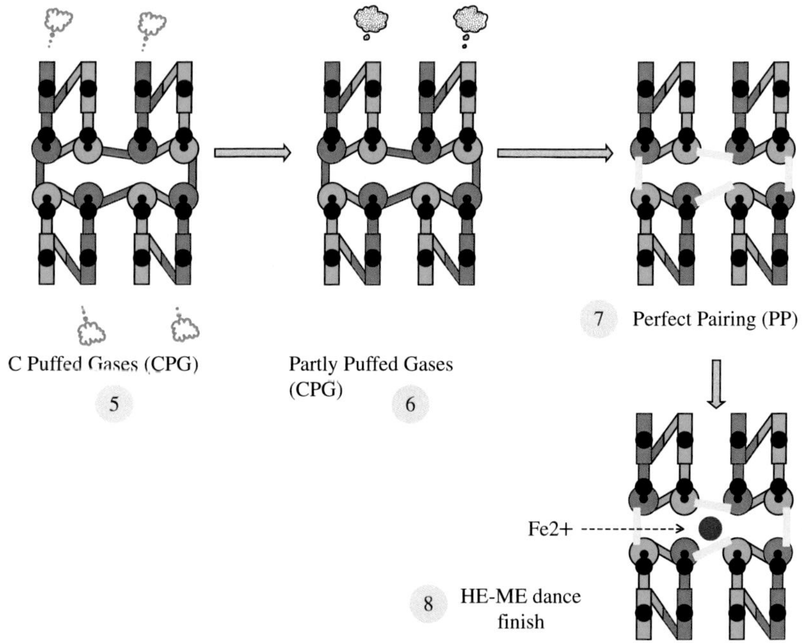

Figure 8.3 Heme dance, Part 2.

HEME SYNTHESIS (TABLE 8.1, FIG. 8.4)

Table 8.1 Correlations Between the He-Me Dance and Heme Synthesis

INSTRUCTIONS	WHAT CORRELATES IN HEME SYNTHESIS PATHWAY
1. SuCoA	Succinyl-CoA
2. A Linear Alignment	ALA formation by Succinyl-CoA and Glycine
3. Please Bind Greens	Porphobilinogen (PBG) formation
4. How Many Bound	Hydroxymethylbilane (HMB) formation
5. U Please Gather	Uroporphyrinogen (UPG)
6. C (see) Puffed Gases	Coproporphyrinogen (CPG) formation by decarboxylation
7. Partly Puffed Gases	Protoporphyrinogen (PPG) formation by decarboxylation and oxidation
8. Perfect Pairing	Protoporphyrin (PP)
9. He-Me	Heme formation

LEARNING HEME SYNTHESIS REACTIONS

Another way to remember heme synthesis reactions is the following:
Draw a heme structure with 8 corners as 8 reactions takes place during heme synthesis.

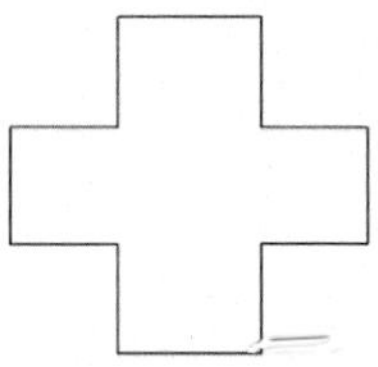

Draw aheme structure with 8 corners as 8 reactions takes place during heme synthesis.

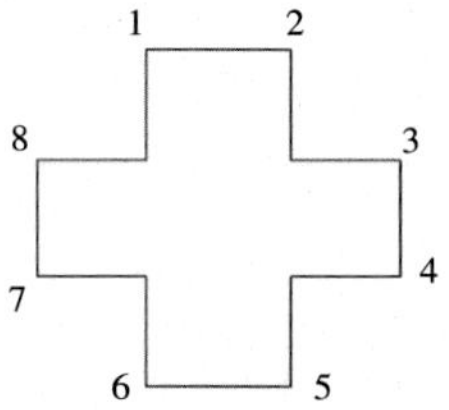

Enumerate the corners as shown

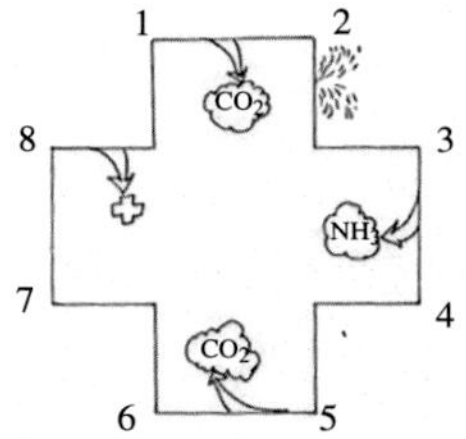

Draw CO_2 exiting from opposite sites as shown above and H_2O splashing afger the 2nd corner. NH_3 released from the right arm. Make heme fall after the 8th

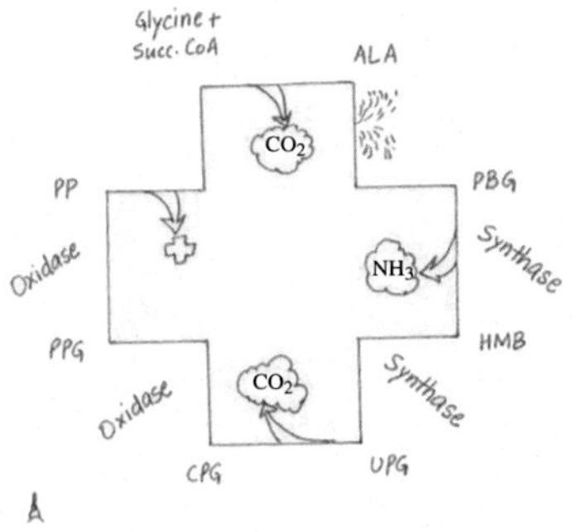

Draw CO_2 exiting from opposite sites as shown above and H_2O splashing afger the 2nd corner. NH_3 is released from the right arm. Make heme fall after the 8th

Figure 8.4 Second trick to learn the heme synthesis reactions.

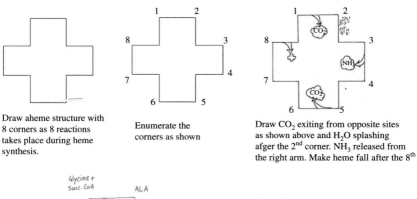

CHAPTER 9

Porphyrias

TRADITIONAL PORPHYRIAS RECAP

Porphyrias are inborn errors of metabolism caused by a deficiency of enzymes involved in heme synthesis (Figs. 9.1 and 9.2). A characteristic feature of porphyrias is the excretion of porphyrins and porphyrinogens in urine. Autosomal dominant inheritance is observed in porphyrias, except ALA Dehydratse-deficiency porphyria and congenital erythropoietic porphyria. Heme synthesis takes place in erythroblasts and the liver, therefore these tissues are affected in porphyrias. Porphyrias are categorized on the basis of the tissue affected in three groups: (1) hepatic porphyria; (2) erythropoietic porphyria; and (3) porphyrias with both hepatic and erythropoietic features.

ALA Synthase ALAS deficiency (erythroid form) results in anemia with low hemoglobin and erythrocyte counts.

ALA Dehydratase ALAD-deficiency porphyria presents with abdominal pain and neuropsychiatric symptoms, along with increased ALA and coproporphyrin III.

Uroporphyrinogen I (UPG I) synthase deficiency leads to acute intermittent porphyria (hepatic form), which is characterized by increased porphobilinogen (PBG) and d-ALA, along with abdominal pain and neuropsychiatric features.

Uroporphyrinogen III (UPG III) synthase deficiency causes congenital erythropoietic porphyria, in which the patient complains of photosensitivity due to elevated uroporphyrin levels in the skin which are photoactive. Uroporphyrins demonstrate a strong red fluorescence under UV light.

Uroporphyrinogen III decarboxylase deficiency porphyria is called porphyria cutanea tarda, and presents with photosensitivity and increased uroporphyrin.

Sweet Biochemistry: Remembering Structures, Cycles, and Pathways by Mnemonics
DOI: http://dx.doi.org/10.1016/B978-0-12-814453-4.00009-1

CPG oxidase deficiency or hereditary coproporphyria includes photosensitivity, abdominal pain, and neuropsychiatric features.

PPG oxidase or variegate porphyria resembles CPG oxidase, but the difference is that in CPG oxidase deficiency, coproporphyrin III is raised, but in PPG oxidase deficiency protoporphyrin IX is increased.

In ferrochelatase deficiency or protoporphyria, protoporphyrin IX is elevated along with photosensitivity.

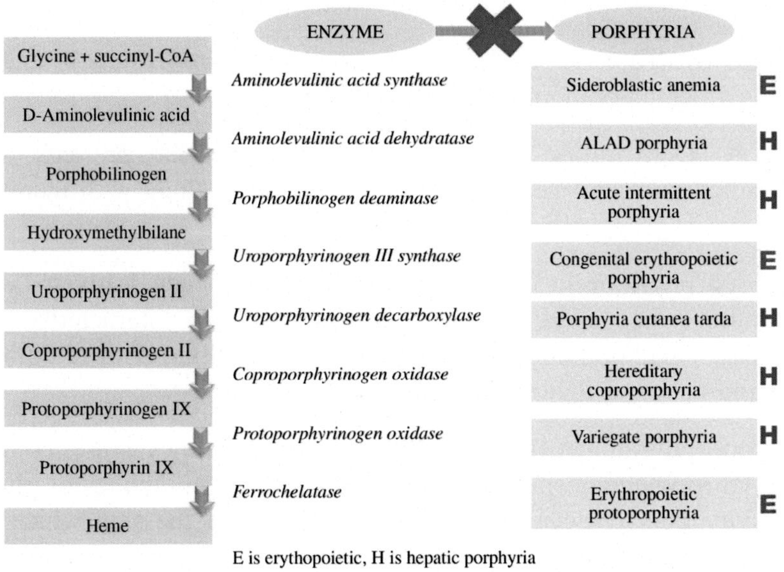

Figure 9.1 Porphyria and defective enzymes.

PORPHYRIA TYPES MNEMONICS

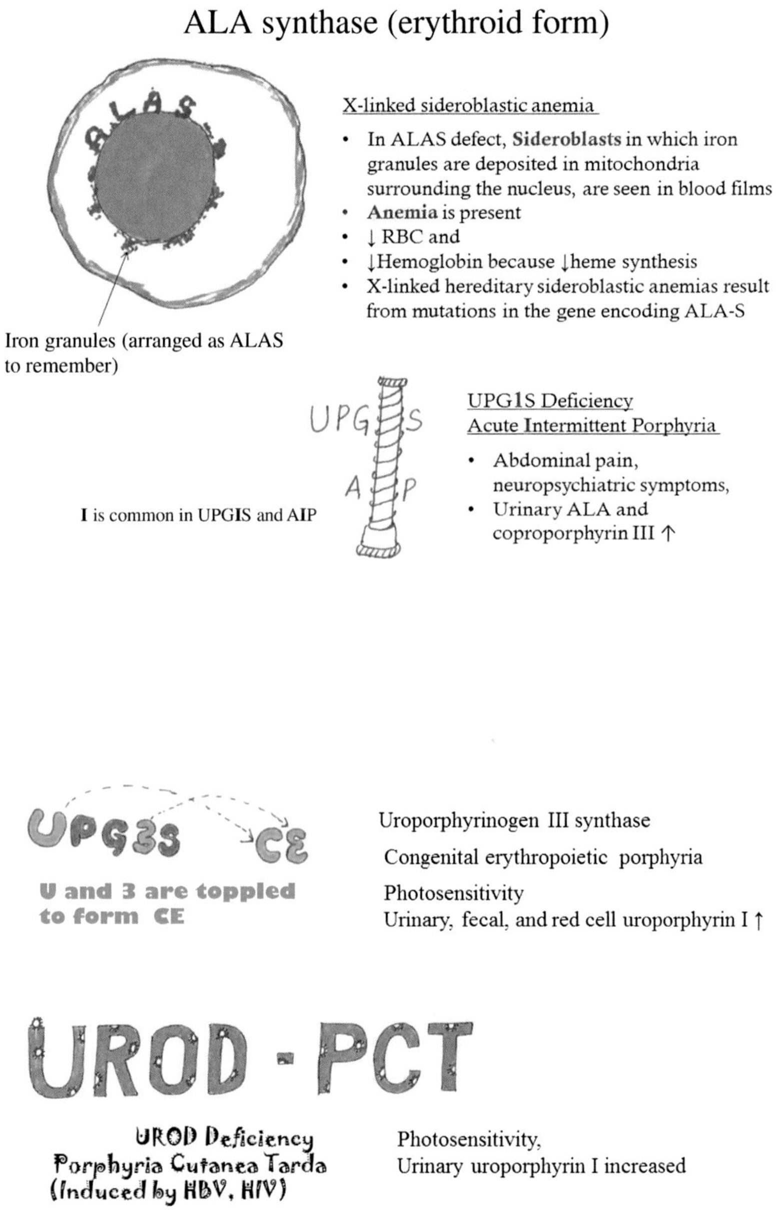
ALA synthase (erythroid form)
ALAS
X-linked sideroblastic anemia
• In ALAS defect, **Sideroblasts** in which iron granules are deposited in mitochondria surrounding the nucleus, are seen in blood films
• **Anemia** is present
• ↓ RBC and
• ↓Hemoglobin because ↓heme synthesis
• X-linked hereditary sideroblastic anemias result from mutations in the gene encoding ALA-S
Iron granules (arranged as ALAS to remember)
UPGIS
AIP
I is common in UPGIS and AIP
UPG1S Deficiency
Acute Intermittent Porphyria
• Abdominal pain, neuropsychiatric symptoms,
• Urinary ALA and coproporphyrin III ↑
UPG3S CE
U and 3 are toppled to form CE
Uroporphyrinogen III synthase
Congenital erythropoietic porphyria
Photosensitivity
Urinary, fecal, and red cell uroporphyrin I ↑
UROD - PCT
UROD Deficiency
Porphyria Cutanea Tarda
(Induced by HBV, HIV)
Photosensitivity,
Urinary uroporphyrin I increased

Figure 9.2 Summary of the porphyrias.

Read HCP and CPOX which indicate hereditary coproporphyria and coproporphyrinogen oxidase

Coproporphyrinogen oxidase (CPOX) deficiency
Hereditary coproporphyria

Photosensitivity, abdominal pain, neuropsychiatric symptoms
Urinary ALA, PBG, and coproporphyrin III and fecal coproporphyrin III increased

See the variegated appearance of V and PPOX is kept in this V

Protoporphyrinogen oxidase (PPOX) deficiency
Variegate porphyria

Photosensitivity, abdominal pain, neuropsychiatric symptoms
Urinary ALA, PBG, and coproporphyrin III and fecal protoporphyrin IX increased

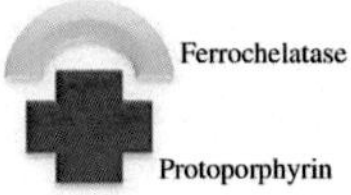

Ferrochelatase deficiency or protoporphyria can be easily remembered as protoporphyrins are the substrates of enzyme ferrochelatase

PORPHYRIA SUMMARY

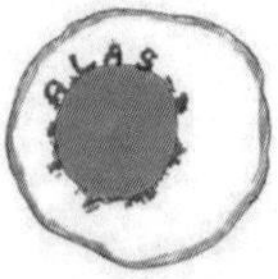

Note the ALAS in sideroblasts

UPG **IS** and AIP share same **I**

U and 3 are toppled to CE

See the virus-infected UROD and PCT

Read the HCP and then CPOX in these letters

Look at the variegated V carrying PPOX

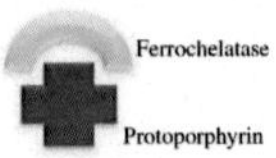

Ferrochelatase is still working on protoporphyrins

CHAPTER 10

Urea Synthesis

TRADITIONAL UREA CYCLE RECAP

The urea or ornithine cycle helps to excrete two harmful gases, ammonia and carbon dioxide, from the body. Ammonia is highly toxic to the central nervous system and needs to be eliminated from the body. The steps of this cycle take place in the mitochondria and cytoplasm. Liver is the main organ synthesizing urea, along with the kidneys to a lesser extent. Alpha nitrogen of amino acids is excreted in the form of urea.

Amino acids transfer their amino groups to alpha ketoacids, like alpha-ketoglutarate (α-KG), by a transamination reaction catalyzed by transaminases with pyridoxalphosphate as coenzyme. Alpha-KG converts to glutamate during this process. Glutamate dehydrogenase acts upon glutamate and releases ammonia along with α-KG.

Chemically urea is $H_2N{-}CO{-}NH_2$. One amino group comes from NH_4^+ and the second from aspartate, while carbon is derived from bicarbonate. The first step in the urea cycle is condensation of CO_2 and NH_3 to carbamoyl phosphate by carbamoyl synthetase 1 (CPS 1), utilizing 2ATP in mitochondria. Carbamoyl phosphate is a high-energy compound. CPS1 is activated by *N*-acetylglutamate and is different from CPS II, which participates in pyrimidine synthesis. *N*-acetyl-glutamate is synthesized by N-acetyl-glutamate synthetase, which is activated by arginine and glutamate.

Ornithine enters mitochondria through a transporter. Ornithine transcarbamoylase (OTC) transfers the carbamoyl group to ornithine, yielding citrulline, which leaves the mitochondria for further steps. L-Aspartate is condensed with citrulline to form argininosuccinate by argininosuccinate synthetase contributing to the second nitrogen of the urea molecule. This step requires ATP and releases AMP. Argininosuccinate is lysed by argininosuccinase to release arginine and fumarate. Urea is produced by hydrolysis of the guanidino group of arginine, along with regeneration of ornithine. Four ATP units are consumed in the synthesis of one urea molecule.

Sweet Biochemistry: Remembering Structures, Cycles, and Pathways by Mnemonics
DOI: http://dx.doi.org/10.1016/B978-0-12-814453-4.00010-8

The urea cycle is linked to the TCA cycle as aspartate donates one amino group of urea and fumarate released during the urea cycle can enter the TCA cycle (Figs. 10.1 and 10.2).

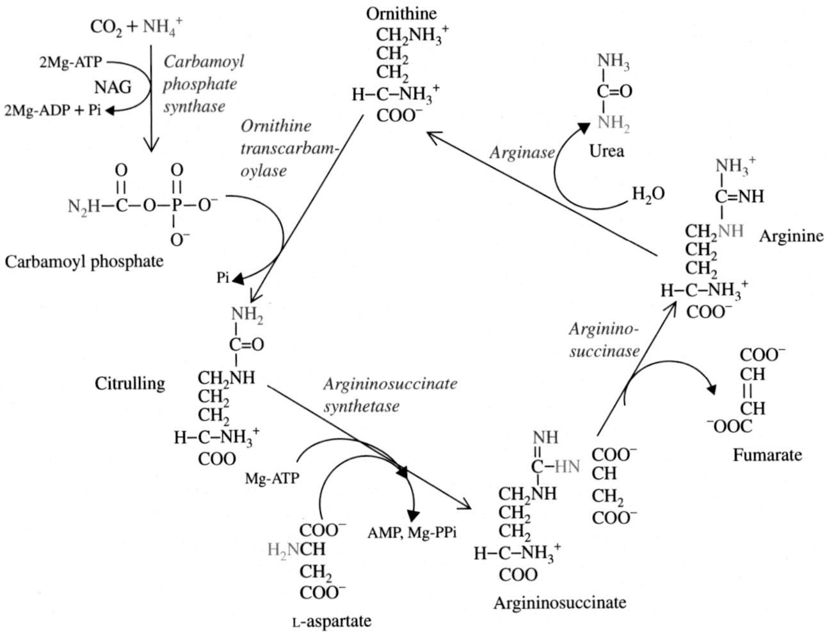

Figure 10.1 Steps of the urea cycle.

UREA CYCLE STEPS

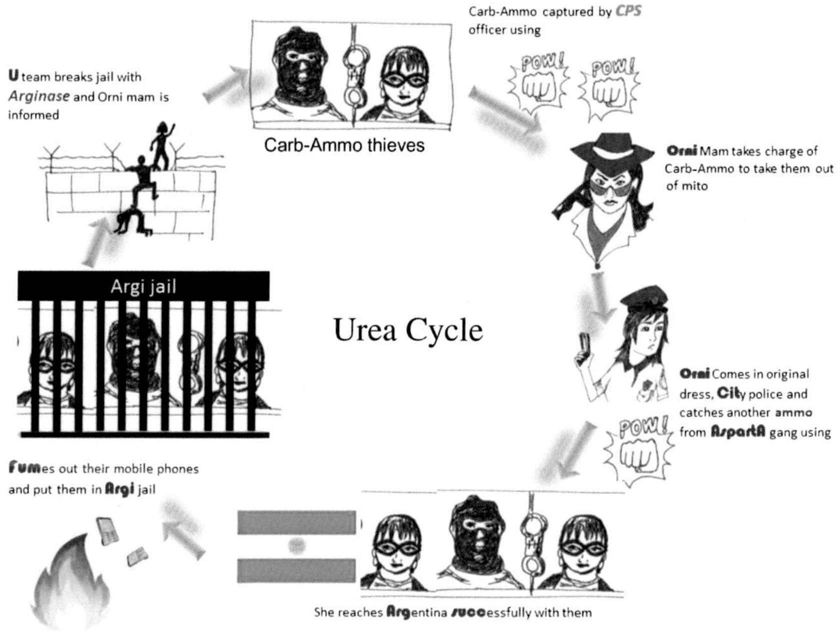

Figure 10.2 Story of bad thieves: Carb and Ammo—the urea cycle.

UREA CYCLE: STORY OF TWO BAD GASES

Two bad thieves: Carb and Ammo were causing a great deal of nuisance in the holy city of Human by forming a secret team carbamoyl. An honest CPS officer captured the two using a lot of energy. Orni Mam takes the two into custody to cautiously take them out of mito (mitochondria).

Orni Mam comes in her original uniform of the City Police. She catches another sister thief Ammo from the asparta gang after a fight. She reaches Argentina successfully with them. Here she takes out their phones and destroys them into fumes so that they cannot contact anyone. The three thieves were sent to Argi jail. Inside jail they formed a U team and escaped the jail using Arginase gel. Orni Mam is again informed about the same, and thus the escape and capture story continues.

UREA CYCLE STORY (TABLE 10.1)

Table 10.1 Correlation of the Two Bad Gases Story and the Urea Cycle

What happens in the story	What this correlates with in the urea cycle
Carb-Ammo	Carbon-dioxide and ammonia
CPS officer	Carbamoyl phosphate synthetase
	ATP
Orni Mam	Ornithine
*Cit*y Police	Citrulline
Asparta gang	Aspartate
*Arg*entina *succ*essfully	Argininosuccinate
*Fum*es out	Fumarate
Argi jail	Arginine
U team	Urea

CHAPTER 11

Urea Cycle Disorders

TRADITIONAL UREA CYCLE DISORDERS

Urea cycle disorders (UCDs) are related to defects of enzymes involved in the urea cycle, and affect all age groups.

All enzymes in this pathway can give rise to UCD. Carbamoyl synthetase I (CPS I), which is found in mitochondria, is used in the urea cycle. The liver is the main site of the urea cycle, with the kidney making a minor contribution. CPS I catalyzes the rate-limiting step. N-acetyl glutamate (NAG) is an allosteric activator of CPS I and is synthesized from acetyl-CoA and glutamate. In high concentrations of ammonia, alpha-ketoglutarate is converted to glutamate and hence is not available for the Krebs cycle. Therefore, ATP production is reduced by an impaired Krebs cycle. This accounts for the highly toxic features of ammonia, such as irreversible brain damage, coma, and death.

Hyperammonemia, encephalopathy, and respiratory alkalosis are the principal signs and symptoms of urea cycle disorders. A characteristic feature of UCD is that ammonia toxicity is highest when the blockage is in the first or second stage. Citrulline products are comparatively less toxic and, moreover, some ammonia has been consumed upto this reaction, hence symptoms are most severe in hyperammonemia I. Ammonia and glutamine are elevated in deficiencies of CPS, OTC, ASS, and ASL.

Patients with UCD have complaints including vomiting, lethargy, irritability, severe mental retardation, and aversion to high-protein foods. Neonates may also suffer from apnea and hypothermia when exposed to high concentrations of ammonia. Members of urea cycle disorders have similar presenting features but at level of enzyme block the substrates of enzyme increases and the concentration of products falls. Transporters, such as ornithine transporters and citrin deficiency, are included in the urea cycle disorders.

Sweet Biochemistry: Remembering Structures, Cycles, and Pathways by Mnemonics
DOI: http://dx.doi.org/10.1016/B978-0-12-814453-4.00011-X

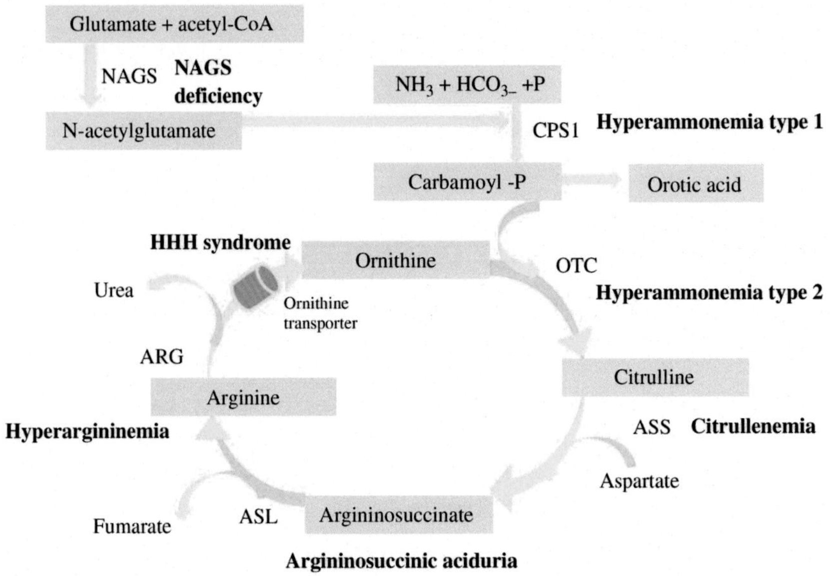

Figure 11.1 Urea cycle and urea cycle disorders.

Infections, high-protein diets, starvation, physical exertion, drugs, and surgery can elevate ammonia levels and aggravate patients' symptoms.

The treatment of urea cycle disorders revolves around decreasing ammonia production by restricting protein intake, which minimizes ammonia level increases and ameliorates brain damage. Levulose is administered to promote an acidic gut environment, which converts NH_3 to NH_4^+, preventing absorption of ammonia from the gut. Compounds like benzoate and phenylacetate are helpful in binding amino acids, and hence lowering ammonia production (Fig. 11.1).

CPSI DEFICIENCY (FIG. 11.2)

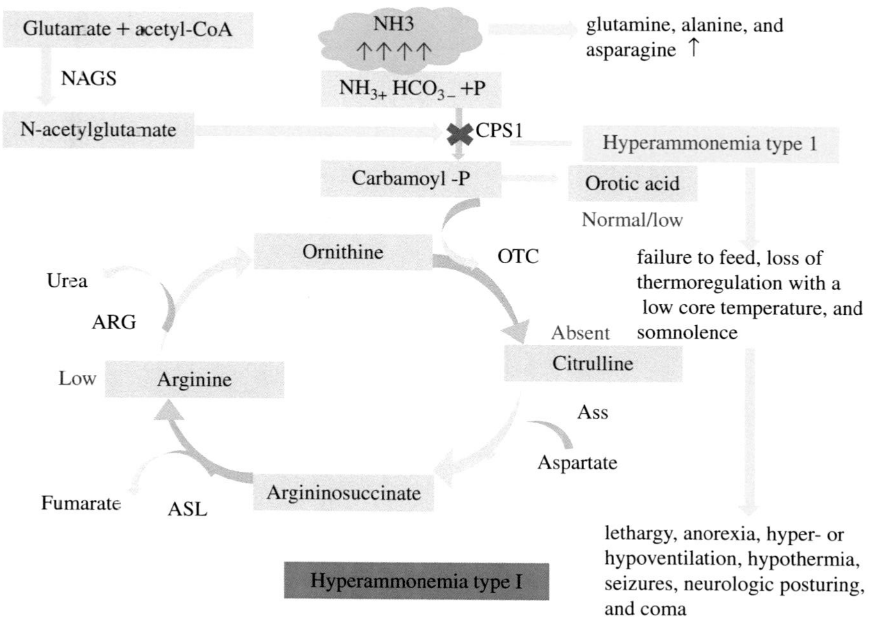

Figure 11.2 CPS deficiency. The yellow arrows indicate the level reactions can reach.

NAGS DEFICIENCY (FIG. 11.3)

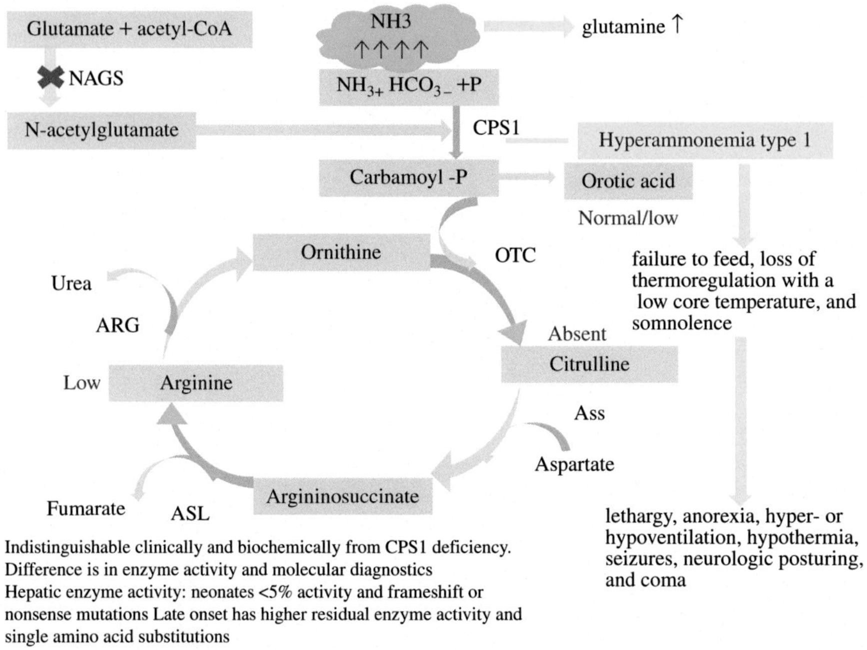

Figure 11.3 NAGS deficiency.

ORNITHINE TRANSPORTER DEFECT (FIG. 11.4)

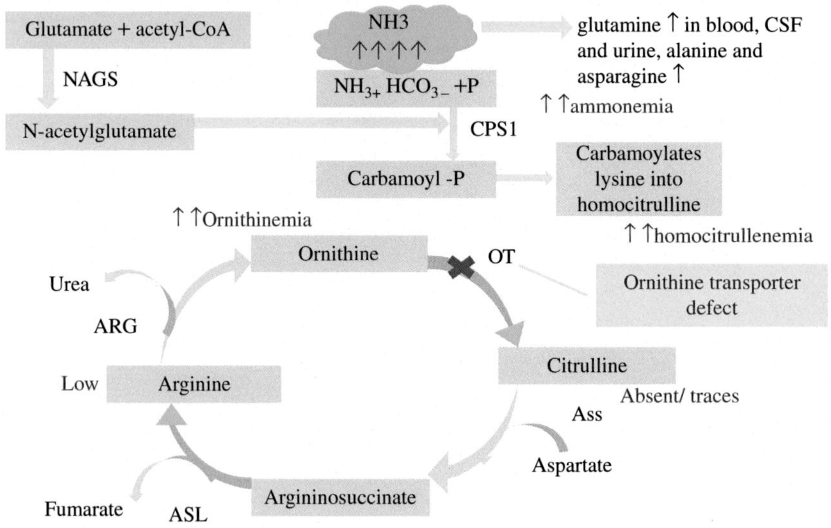

Figure 11.4 Ornithine transporter defect.

ORNITHINE TRANSCARBAMOYLASE DEFICIENCY (FIG. 11.5)

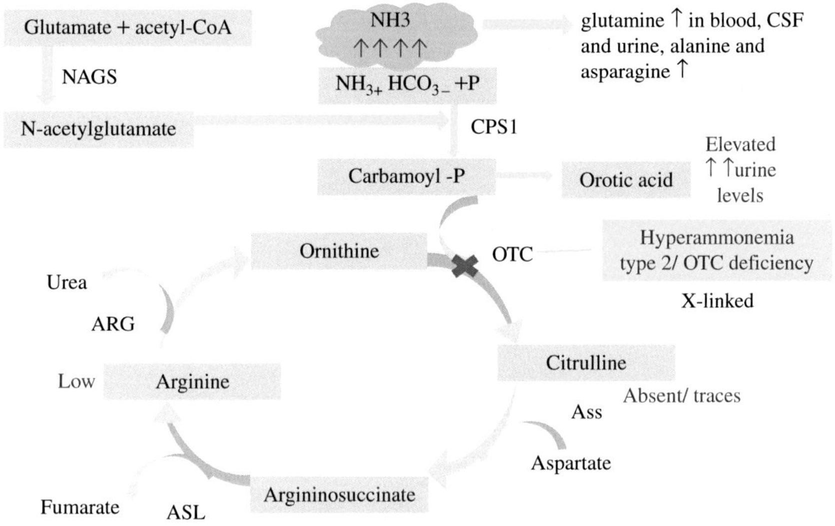

Figure 11.5 Ornithine transcarbamoylase deficiency.

CITRULLENEMIA TYPE 1 (FIG. 11.6)

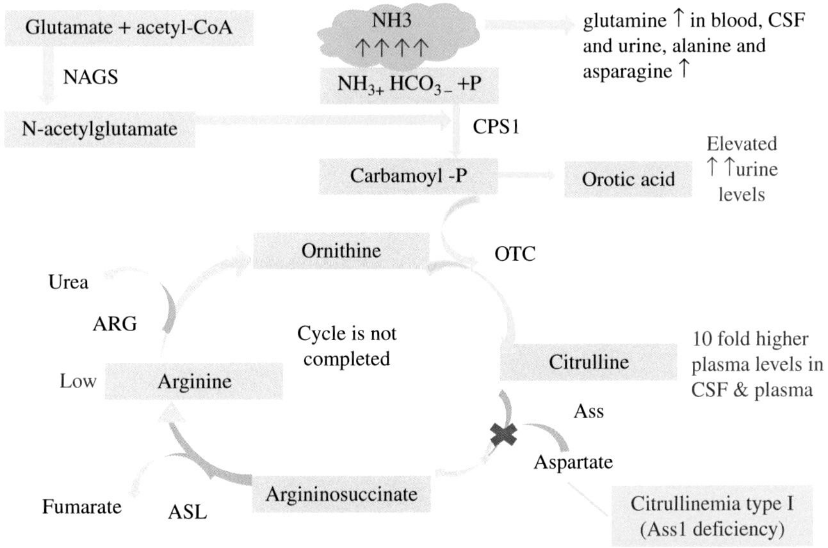

Figure 11.6 Citrullenemia type 1.

ARGININOSUCCINIC ACIDURIA (FIG. 11.7)

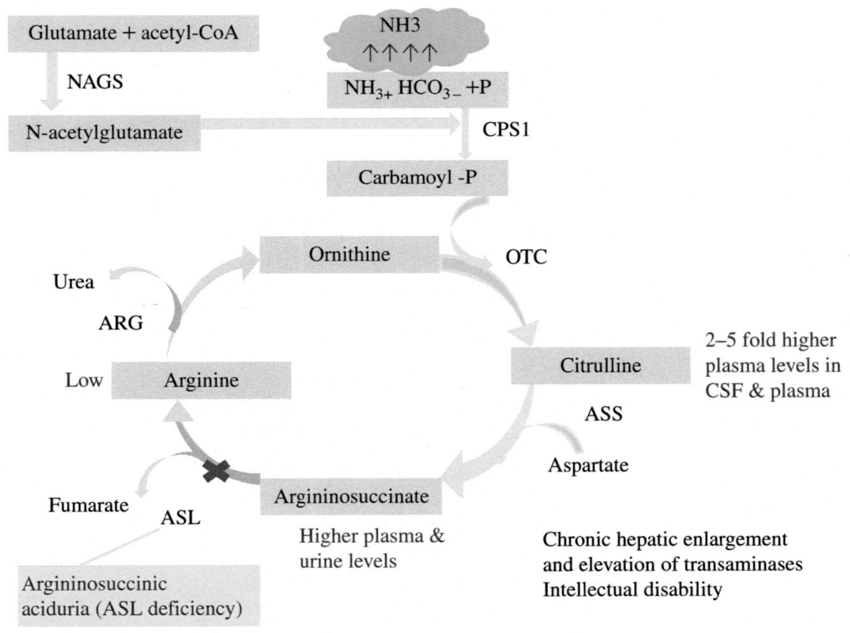

Figure 11.7 Argininosuccinic aciduria.

ARGINASE DEFICIENCY (FIGS. 11.8 AND 11.9)

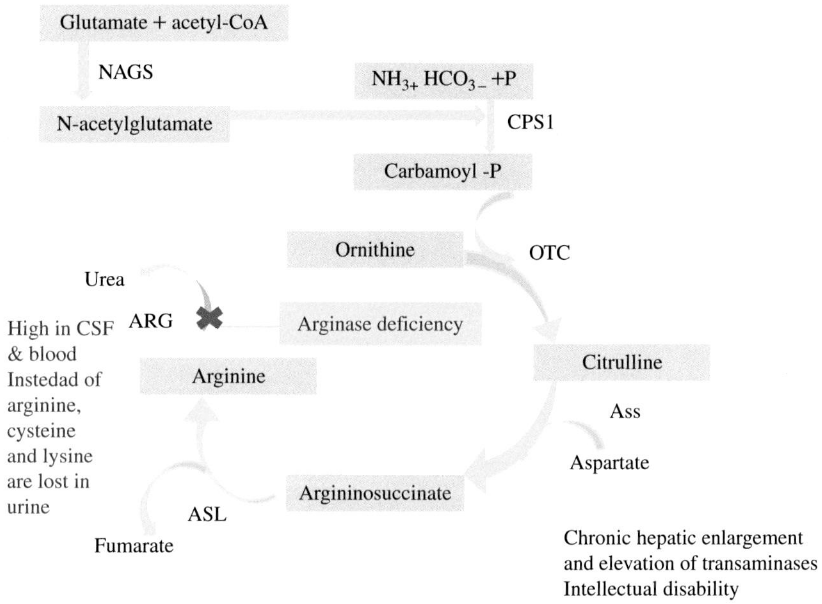

Figure 11.8 Arginase deficiency.

UREA CYCLE DISORDERS SUMMARY

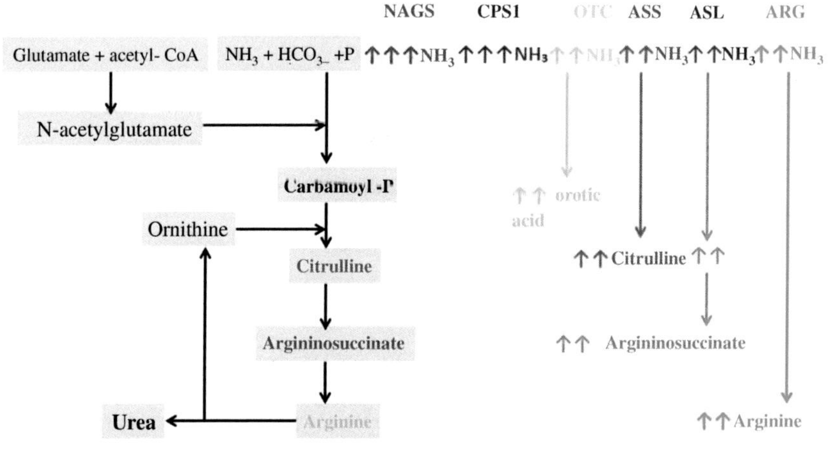

Figure 11.9 Urea cycle disorders summary. At the top of the diagram enzymes are written on the right-hand side. Arrows with the same color lines indicate the level to which the pathway works. The main accumulating substances are listed for each disorder.

CHAPTER 12

Glycogen Storage Disorders

TRADITIONAL GLYCOGEN STORAGE DISORDERS RECAP

Glycogen is a highly branched storage form of glucose in animals. Glucose units are joined by alpha-1,4-glucosidic linkages and branching carries an alpha-1,6-glucosidic bond. The liver and muscle are the two major areas of glycogen storage (muscle being the larger source). Liver, by glycogenolysis, releases free glucose into the blood, while muscle lacks glucose-6-phosphatase, so the glucose-6-P in muscle starts glycolysis and forms pyruvate which causes gluconeogenesis.

Glycogen synthesis: The glycogenin protein core attaches glucose on its tyrosine residue autocatalytically in the initiation of synthesis. UDP-glucose provides glucose for this reaction. Glycogen synthase then forms glucosidic bonds between C-1 of the glucose of UDPGlc and C-4 of the terminal glucose residue of glycogen. The branching enzyme creates a branch with an alpha-1,6 bond, when the chain is 11 glucose units long. This process of lengthening and branching continues, yielding a highly branched glycogen molecule.

Glycogen breakdown: Glycogen degradation starts with the addition of phosphates to glucose from nonreducing ends by glycogen phosphorylase, the rate-limiting enzyme. Glucose-1-P is released sequentially until four glucose residues remain on the branch. Transferase shifts trisaccharide to the other branch, leaving behind a glucose joined with an alpha-1,6 bond. Debranching enzymes hydrolyze the alpha-1,6-glycosidic bond to release free glucose. Glycogen phosphorylase continues its action on the straight chain.

Glycogen synthase and glycogen phosphorylase are regulated by phosphorylation. cAMP, which is increased by a number of factors, activates cAMP dependent protein kinase. This enzyme phosphorylates

Sweet Biochemistry: Remembering Structures, Cycles, and Pathways by Mnemonics
DOI: http://dx.doi.org/10.1016/B978-0-12-814453-4.00012-1

phosphorylase kinase and activates it. Phosphorylase kinase in turn activates glycogen phosphorylase.

Glycogen storage diseases are also known as glycogenosis or dextrinosis. They are a group of inherited disorders resulting from defective glycogen synthesis or degradation, leading to accumulation of glycogen in the liver, muscles, and other tissues. Deposited glycogen may be either normal or abnormal in structure. Glycogen accumulation causes hepatomegaly (kidney enlargement) in some, hypoglycemia, and muscle weakness. Glucose-6-phosphate from muscles enters the Pentose Phosphate Pathway (PPP) pathway to elevate purine synthesis, leading to hyperuricemia (Fig. 12.1).

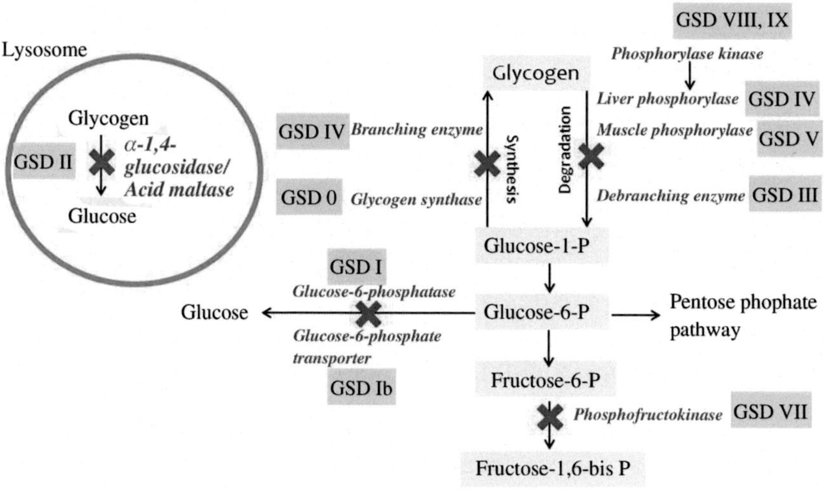

Figure 12.1 Glycogen storage disorders along with enzyme blocks.

GLYCOGEN STORAGE DISORDERS SUMMARY (TABLES 12.1 AND 12.2, FIG. 12.2)

Table 12.1 Summary of Glycogen Storage Disorders with Enzymes and Clinical Features

Type	Enzyme defect	Clinical features
Type 0	Liver glycogen synthase	Hypoglycemia, hyperketonemia, mild hepatomegaly, early death
Type Ia (Von Gierke's disease)	G-6-phosphatase deficiency	Hypoglycemia, massive hepatomegaly, renomegaly, hyperuricemia, ketosis, lactic acidosis, hyperlipidemia, and short stature
Type Ib	Endoplasmic reticulum G-6- P transporter defect	Type Ia features with doll-like faces with fat cheeks, short stature, and protuberant abdomen, bleeding tendency, recurrent infections
Type II (Pompe's disease)	Lysosomal acid maltase	Lysosomes laden with glycogen, diminished myscle tone, cardiorespiratory failure, enlarged tongue
Type III (Cori's disease)	Liver and muscle debranching enzyme	Hypoglycemia, cirrhosis, hepatomegaly, muscle and cardiac involvement (milder course than type I), characteristic branched polysaccharide deposits
Type IV (Anderson's disease)	Branching enzyme	Hepatosplenomegaly, liver cirrhosis, failure to thrive, death usually before 5 years, glycogen with few branch points deposited
Type V (McArdle syndrome	Muscles phosphorylase	Exercise intolerance, myalgia (muscle pain), muscle contracture, hyperCKemia and myoglobinuria
Type VI (Her's disease)	Liver phosphorylase	Hepatomegaly, hypoglycemia, ketosis, and moderate growth retardation
Type VII (Tarui's disease)	Muscle, RBC PhosphoFructoKinase 1	Muscle pain and cramps, decreased endurance
Type VIII	Liver phosphorylase kinase	Mild hepatomegaly, mild hypoglycaemia
Type IX	Liver and muscle phosphorylase kinase	Hepatomegaly, mild hypoglycaemia, short stature, muscle weakness
Type IX	Liver and muscle phosphorylase kinase	Hepatomegaly, mild hypoglycaemia, short stature, muscle weakness
Type X	cAMP dependent Protein kinase A	Muscle aches or cramping following strenuous physical activity, recurrent myoglobinuria, kidney failure

Table 12.2 Correlation of the Story with Glycogen Storage Disorders

Type	Story Part Correlating	Disease	Enzyme Defect
0	It's a Great Story	Glycogen synthase defect	Glycogen Synthase
1a	Von (one) Gierke was a Great 6 Pack abs warrier	Von Gierke's disease	G-6-Ptase deficiency
1b	and transported		ER G-6-P Transporter defect
II	a pompe with acid maltase	Pompe's disease	Lysosomal Acid maltase
III	He adored Cori—a debrancher	Cori's disease	Liver and muscle debranching enzyme
IV	While he was a brancher from Andery	Anderson's disease	Branching enzyme
V	He fought with McArdle with large muscle phoscle	McArdle syndrome	Muscle. Phosphorylase
VI	to win Her as a life partner	Her's disease	Liver Phosphorylase
VII	On the day of Tarui, with blood (RBC) stained muscles and one PFK sword	Tarui's disease	Muscle, RBC PhosphoFructoKinase 1
VIII	He suffered 8 liver powerful kuts (cuts)		Liver Phosphorylase Kinase
IX	And 9 muscle phoscle kuts (cuts)		Liver and Muscle Phosphorylase Kinase
X	Finally impressed, she married him in camp of deproka.		cAMP dependent Protein Kinase A

STORY MNEMONIC OF GLYCOGEN STORAGE DISORDERS

Glycogen storage disorders are very difficult to remember. I have represented this in the form of commonest warrior story. First enjoy the story in Figure 12.2 and then try to remember it word by word. Correlated the words of story with Table 12.2.

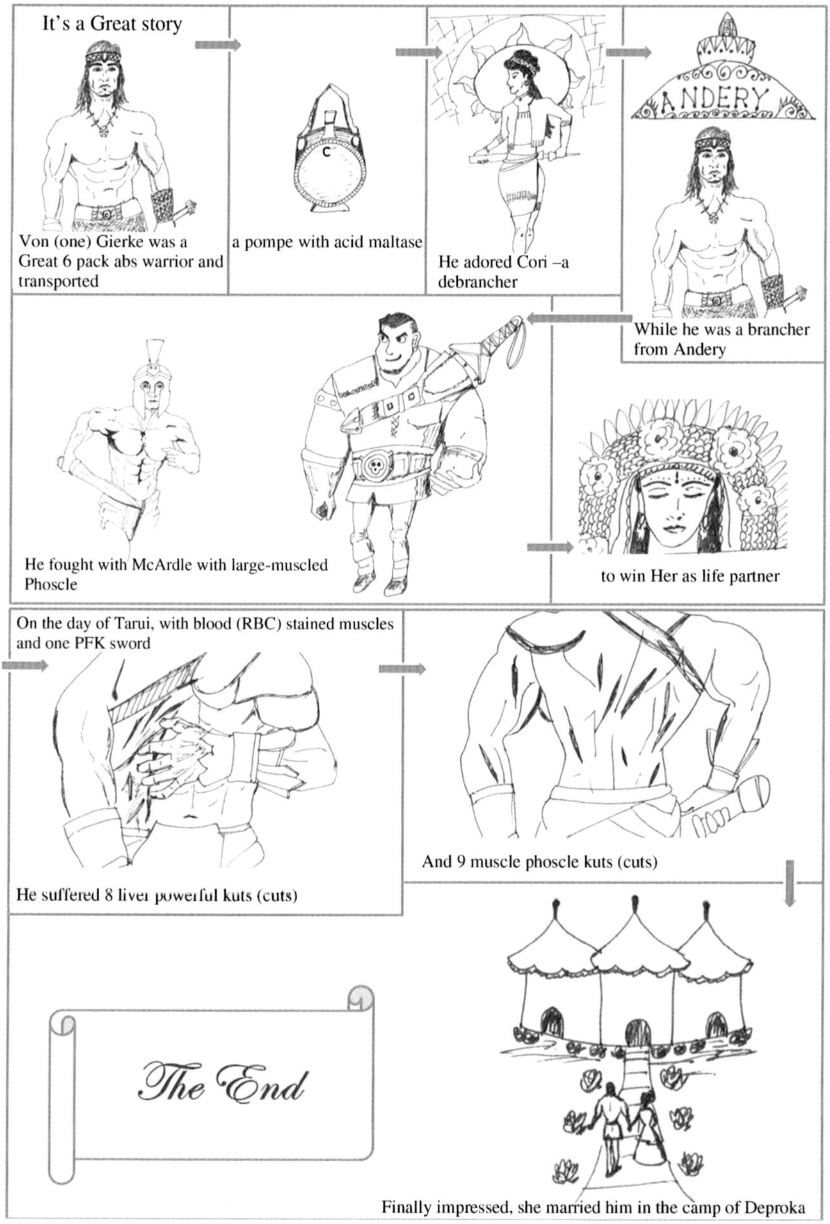

Figure 12.2 Glycogen storage disorders depicted as the story of Von Gierke.

CHAPTER 13

Ceramide Structure and Derivatives

CERAMIDE AND SPHINGOSINE STRUCTURE (FIG. 13.1)

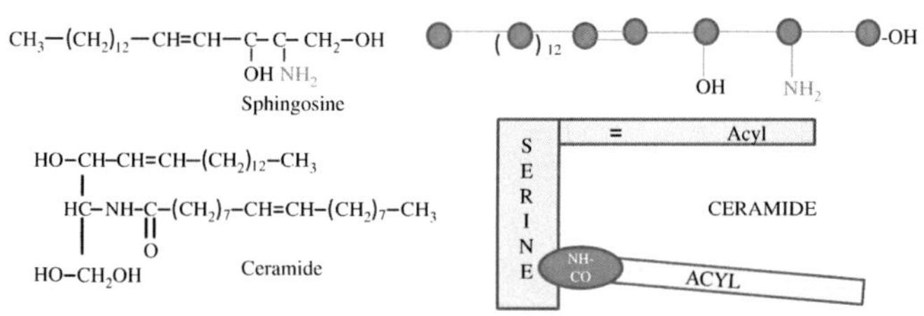

Sphingosine is an 18-carbon amino alcohol with a long unsaturated hydrocarbon chain.

Ceramide is a longchain fatty acid amide derivative of sphingosine or, simply it is the combination of sphingosine alcohal wiht a fatty acid. Derivatives of ceramide are formed by attaching an OH group to C1.

Ceramide plays an important role in signal transduction in programmed cell death (apoptosis), the cell cycle and cell differentiation and senescence. Ceramide is the core molecule of sphingolipids.

SPHINGOSINE STRUCTURE

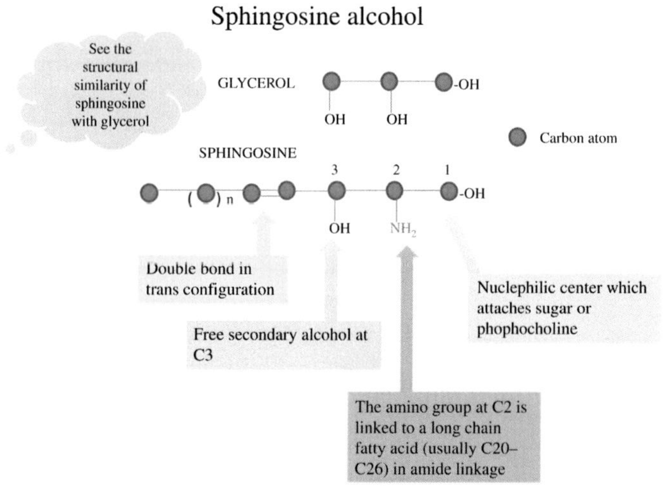

Figure 13.1 Comparison of the structures of glycerol and sphingosine.

Sweet Biochemistry: Remembering Structures, Cycles, and Pathways by Mnemonics
DOI: http://dx.doi.org/10.1016/B978-0-12-814453-4.00013-3

CERAMIDE STRUCTURE DESCRIPTION (FIG. 13.2)

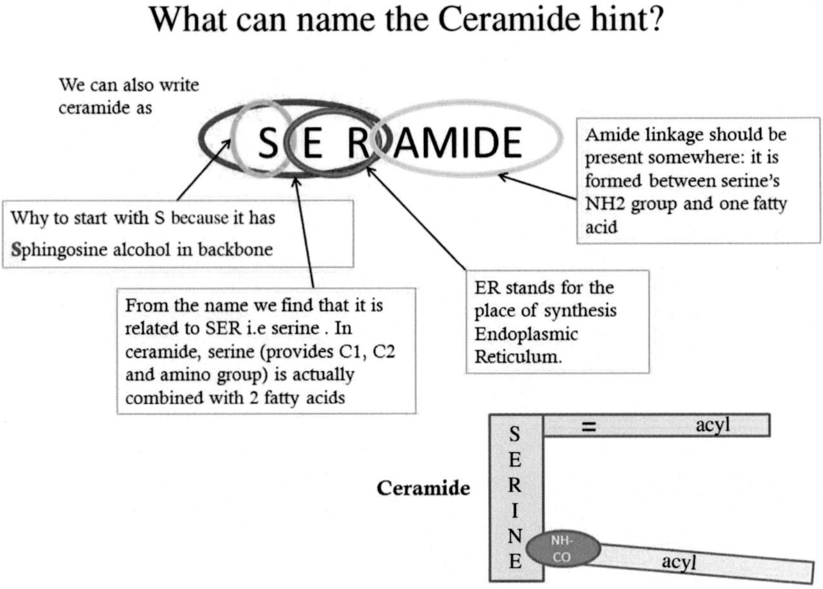

Figure 13.2 Ceramide structure description.

CERAMIDE DERIVATIVES

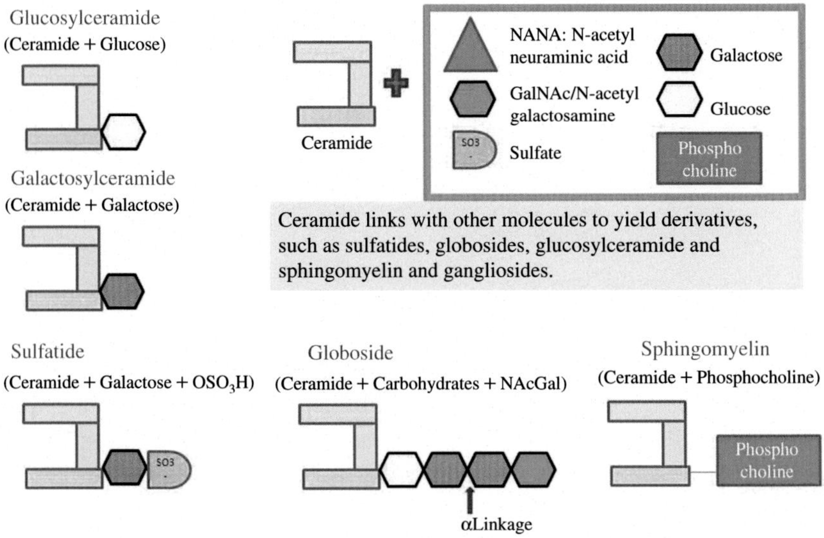

Galactocerebroside and **glucocerebroside** are ceramide linked with galactose or glucose. These carbohydrates are joined with the help of activated nucleotide sugars UDPgalactose and UDPglucose.

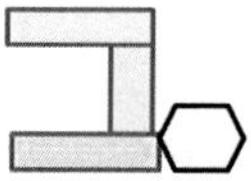

Sulfatide or sulfogalactocerebroside is a sulfuric acid ester of galactocerebroside.

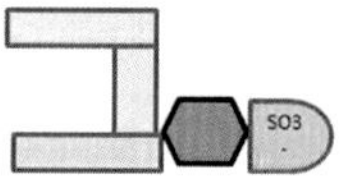

Globosides are cerebrosides having two or more sugar residues, commonly galactose, glucose, or N-acetylgalactosamine (GalNAc)

Sphingomyelin is aceramide phosphocholine and **an important component of the myelin sheath of neurons.** Hydroxyl (-OH) group at C1 of sphingosine is esterified to choline through a phosphodiester bond and the NH_2 group unites with a long-chain fatty acid through an amide bond to form sphingomyelin. Thus sphingomyelin is also a phospholipid. It is neutral at physioloigical pH becaused it has an equal positive and negative charge.

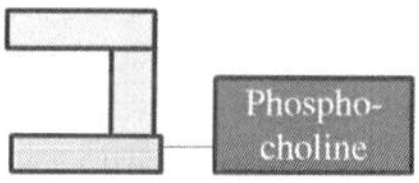

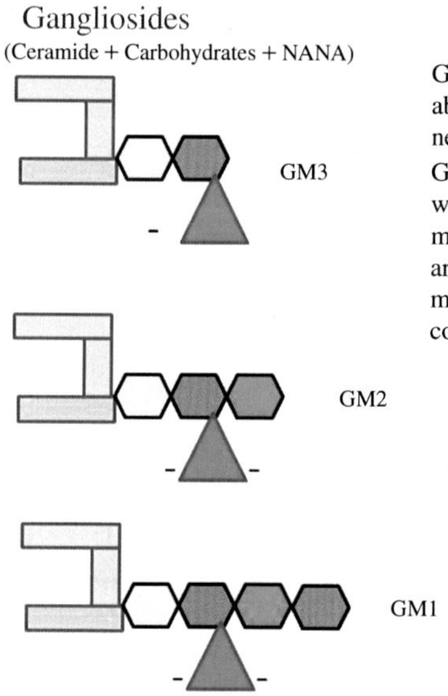

Ganglioside are sialic acid rich glycosphingolipids abundantly present in ganglion cells of the central nervous system.

Gangliosides are written as GM1, GM2, and GM3, where G indicates ganglioside while M stands for monosialo-containing species. Numbers 1, 2 or 3 are assigned according to chromatographic migration. GM3 is the simplest and GM1 is more complex ganglioside.

CHAPTER 14

Lipid Storage Disorders/Sphingolipidoses

TRADITIONAL SPHINGOLIPIDS CATABOLISM

Sphingolipidoses are lysosomal storage disorders in which enzymes of sphingolipid catabolism are absent. In these disorders, the sites of sphingolipid catabolism like lysosomes of phagocytes, histiocytes or macrophages in bone marrow, liver and spleen are affected.

Sphingolipid catabolism is carried out by hydrolases present in lysosomes and the structural components are removed sequentially. The structural components of sphingolipids are described in Chapter 13, Ceramide Structure and Derivatives. The various enzymes participating in this pathway include galactosidases, glucosidases, neuraminidase, hexosaminidase, sphingomyelinase (a phosphodiesterase), sulfatase, and ceramidase (an amidase). Irreversible reactions catalyzed by these enzymes break down the molecule into its building blocks. These diseases generally affect the pediatric age group.

Other features of sphingolipidoses include the following. Ceramide compound deposition in the central nervous system causing neurodegeneration and mental retardation. Degeneration is diffuse in nature. Enzymes are defective in lysosomal degradation of complex lipids but synthesis is normal. Deposition of substrate of deficient enzyme, is seen.

The main sphingolipidose diseases are Niemann-Pick's disease, Tay-Sach's disease, Gaucher's disease, Fabry's disease, metachromatic leukodystrophy, Farber's disease, and Sandoff's disease. Patients often suffer from psychomotor retardation, weakness, and spasticity. Antenatal diagnosis is available (Fig. 14.1)

Sweet Biochemistry: Remembering Structures, Cycles, and Pathways by Mnemonics
DOI: http://dx.doi.org/10.1016/B978-0-12-814453-4.00014-5

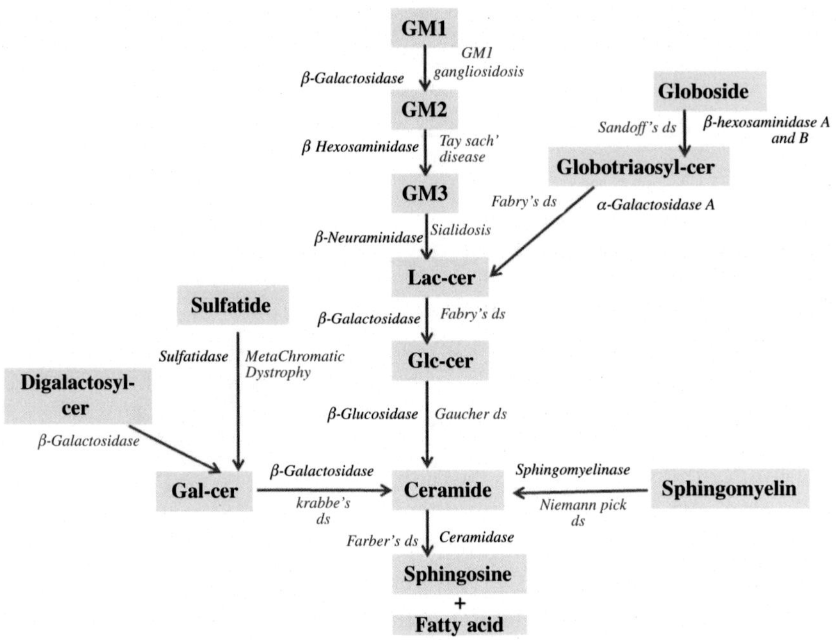

Figure 14.1 Catabolism of sphingolipids along with related disorders.

REMEMBERING TRICK FOR SPHINGOLIPID CATABOLISM

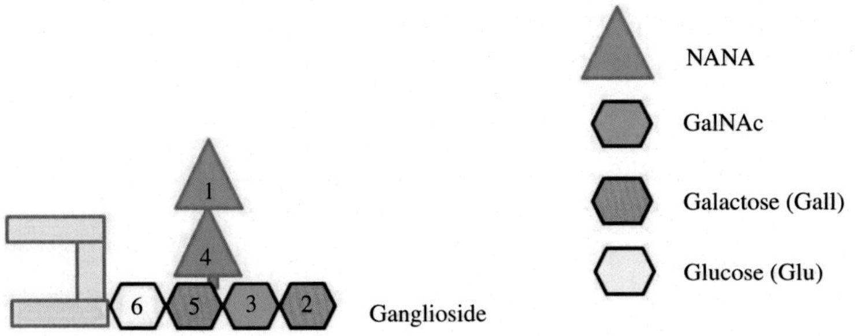

The sequence of removal of components needs to be remembered to memorize the degradation of gangliosides that correlate with the associated diseases. Enzyme names are also related to the substrate, e.g., to remove:

- NANA—neuraminidase is used;
- GalNAc—hexosaminidase is used;
- Gal—galactosidase is used;
- Glu—glucosidase is used.

All carbohydrates are attached by beta linkage, except the first galactose of globoside. Sulfatides, globosides, and sphingomyelin are catabolized by sequentially removing components from the outer ends as shown in Figs. 14.2 and 14.3.

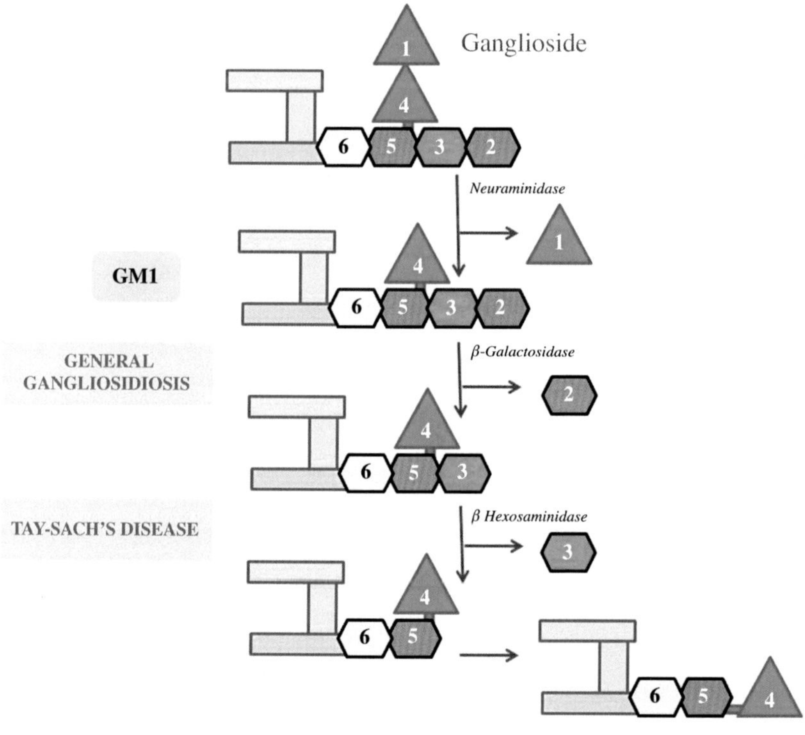

Figure 14.2 Sphingolipid catabolism steps. The sequence of removal of moieties and the enzymes involved can be observed.

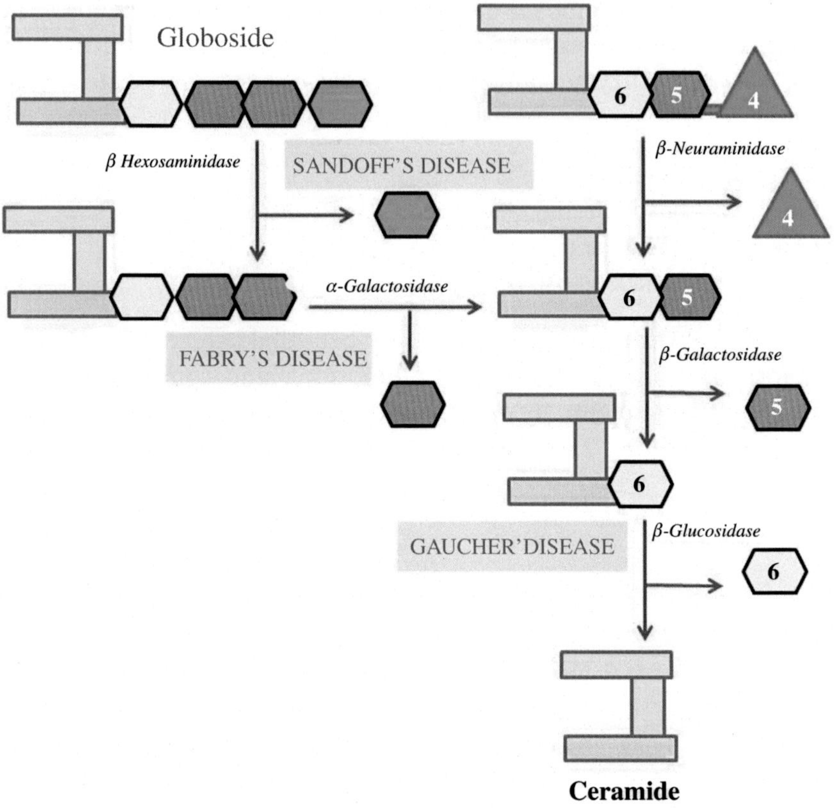

Figure 14.3 Sphingolipid catabolism continued.

TAY-SACH'S DISEASE AND SANDOFF'S DISEASE

Fig. 14.4.

Gangliosidose disorders result from a deficiency of beta-hexosaminidase.

Gangliosides are deposited. GM2 deposition disorders include:

Tay-Sachs disease is also known as GM2 gangliosidosis—variant B. It is caused by a deficiency of hexosaminidase A.

Symptoms begin by 6 months of age and include progressive loss of mental ability, dementia, deafness, difficulty in swallowing, blindness, cherry-red spots in the retinas, and some muscular weakness.

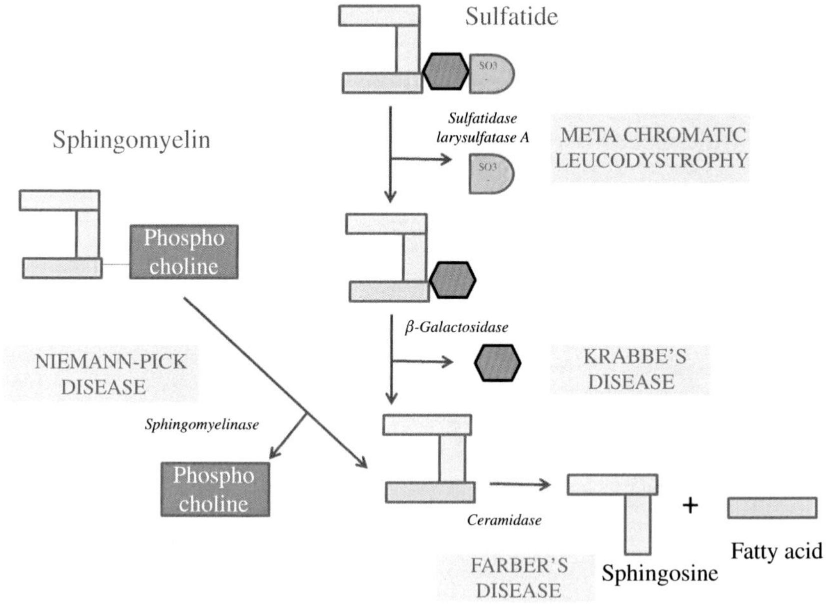

Figure 14.4 Sphingolipid catabolism steps.

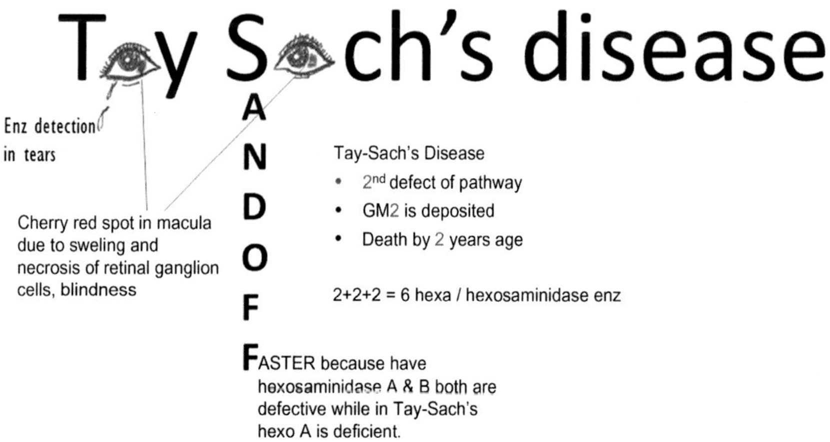

Figure 14.5 Tay-Sach's disease.

Seizures may begin in the second year of life. Anticonvulsant medications are helpful initially

The enzyme is detected in tears.

Sandhoff's disease (variant AB) is a severe form of Tay-Sach's disease.

It is caused by a deficiency of hexosaminidase A and B and onset usually occurs at 6 months of age.

Manifestations of Sandoff's disease resemble Tay-Sach's disease but are faster in onset and are of increased intensity (Fig. 14.5).

GAUCHER'S DISEASE

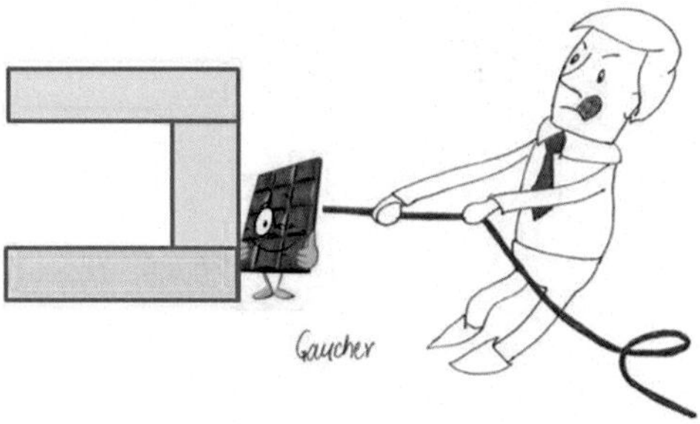

Gaucher is trying to break glucose (chocolate) from ceramide as beta-glucosidase is absent

Figure 14.6 Gaucher's disease.

- Gaucher's disease is caused by a deficiency of glucocerebrosidase.
- Glucosylceramide is deposited in various organs causing symptoms such as enlarged spleen and liver, liver malfunction, bone lesions causing pain and fractures, severe neurologic complications, swelling of lymph nodes, distended abdomen, a brownish tint to the skin, anemia, low blood platelets, and yellow spots in the eyes.
- Susceptibility to infection may increase.
- The disease affects males and females equally (Fig. 14.6).

NIEMANN-PICK'S DISEASE

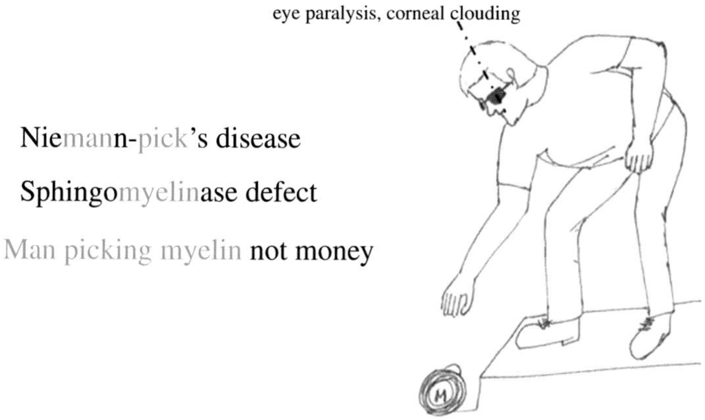

Figure 14.7 Niemann-Pick's disease.

- Niemann-Pick's disease is an autosomal recessive disorder.
- Accumulation of sphingomyelin in cells of the liver, spleen, bone marrow, lungs, and, in some patients, the brain.
- Neurological complications may include ataxia, eye paralysis, brain degeneration, spasticity, loss of muscle tone, and corneal clouding. A characteristic cherry-red halo develops around the center of the retina in 50 percent of patients (Fig. 14.7).

FARBER'S DISEASE

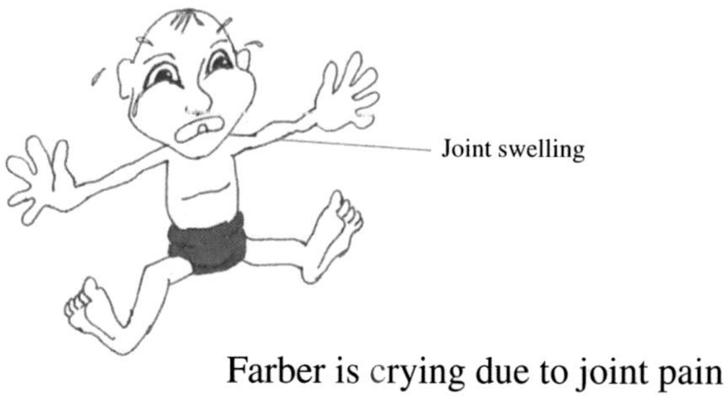

Joint swelling
Farber is crying due to joint pain

Figure 14.8 Farber's disease.

Farber's disease (Fig. 14.8), also known as Farber's lipogranulomatosis, is caused by a deficiency of ceramidase and accumulation of ceramide in the joints, tissues, and central nervous system. It is an autosomal recessive disorder affecting both males and females. Disease onset is typically in early infancy.

Symptoms include moderately impaired mental ability and problems with swallowing. Other symptoms may include joints swelling and contractures, vomiting, arthritis, swollen lymph nodes, with hepatosplenomegaly at birth in the severe form. Death usually occurs by age 2 (Figs. 14.9–14.12).

FABRY'S DISEASE

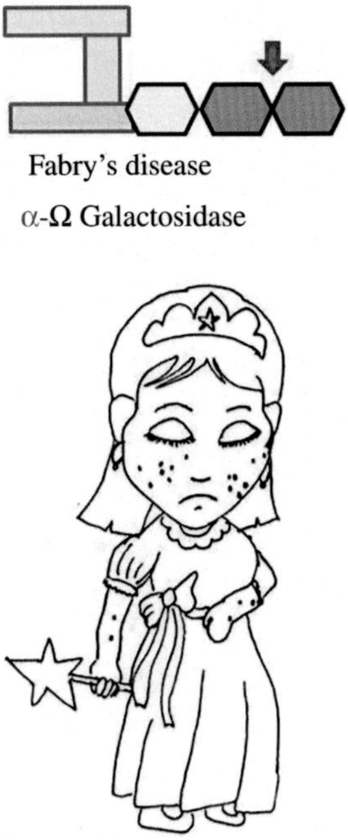

Fabry disease,

- Alpha-galactosidase-A deficiency
- Causes a build-up of globotriaosylceramide in the autonomic nervous system, eyes, kidneys, and cardiovascular system
- Only X-linked lipid storage disease
- Neurological signs include burning pain in the arms and legs
- Risk for stroke or heart attack, heart enlargement`
- Progressive kidney impairement leading to renal failure
- Gastrointestinal difficulties, painful abdomen
- Angiokeratomas (small, non cancerous, reddish-purple elevated spots on the skin)

We assume a fairy girl
As this disease is X-linked.
The fairy is holding her abdomen due to
kidney failure and she is not a very
beautiful fairy because of skin rash.

Figure 14.9 Fabry's disease.

GENERAL GANGLIOSIDOSIS

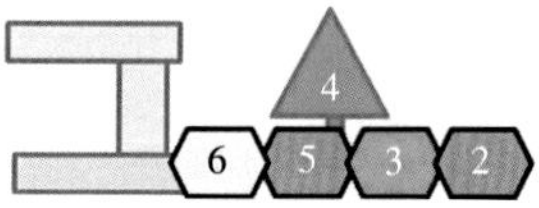

β-Galactosidase deficiency

Here is a general grandpa at his house who is the 1st membere (so GM1 is deposited here and widespread deposition)

GM1 gangliosidoses

- Deficiency of the enzyme beta-galactosidase
- Abnormal storage of GM1 ganglioside particularly in the nerve cells in the central and peripheral nervous systems causing neurodegeneration and seizures
- Liver and spleen enlargement
- Coarsening of facial features
- Skeletal irregularities, muscle weakness
- Joint stiffness
- Cherry-red spots in the eye
- Angiokeratomas

Figure 14.10 General gangliosidosis.

KRABBE'S DISEASE

Krabbe is a keazy kid who has lost his myelin in the school

Krabbe's disease

- Also called globoid cell leukodystrophy and galactosylceramide lipidosis
- Autosomal recessove disorder
- Deficiency of the enzyme galactocerebrosidase
- The disease most often affects infants
- The buildup of galactosylceramide affects the growth of the nerve's protective myelin sheath and causes severe deterioration of mental and motor skills
- Characteristic grouping of cells into globoid bodies in the white matter of the brain, demyelination of nerves and degeneration, and destruction of brain cells

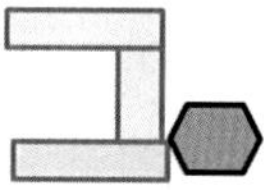

β-Galactosidase defect

Figure 14.11 Krabbe's disease.

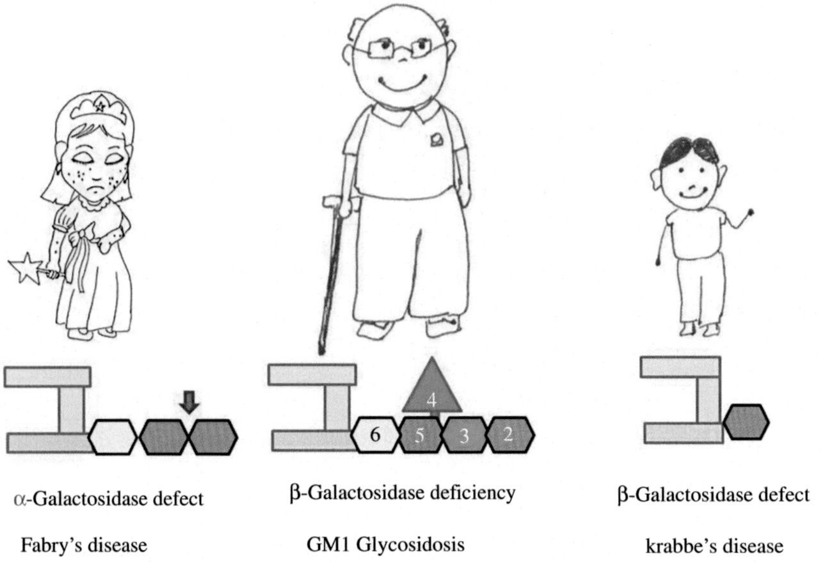

Figure 14.12 Three galactosidase-deficients standing together.

METACHROMATIC LEUKODYSTROPHY (MLD)

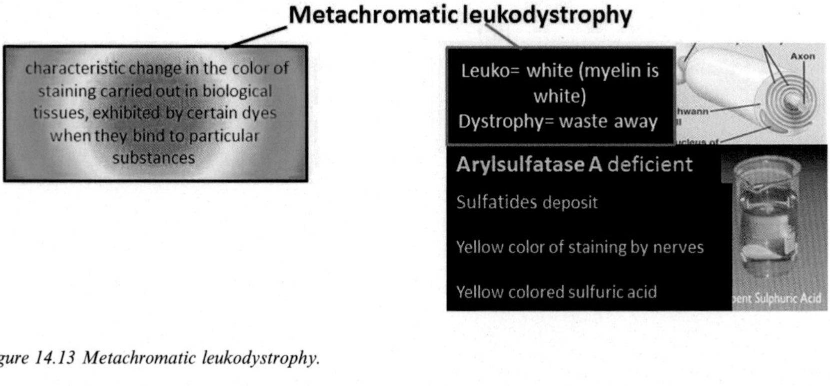

Figure 14.13 Metachromatic leukodystrophy.

MLD is caused by a deficiency of the enzyme arylsulfatase A and is characterized by deposition of 3-sulfogalactosylceramide in the white matter of the central nervous system and in the peripheral nerves and to some extent the kidneys (Fig. 14.13). It is an autosomal recessive disorder in which the myelin sheath is affected.

Onset is typically between 12 and 20 months following birth. Infants may appear normal at first but develop difficulty in walking and a tendency to fall, followed by intermittent pain in the arms and legs, progressive loss of vision leading to blindness, developmental delays, impaired swallowing, convulsions, and dementia before age 2. Death usually occurs before 5 years.

CHAPTER 15

Mucopolysaccharidoses

TRADITIONAL MUCOPOLYSACCHARIDES

Glycosaminoglycans (GAGs) are also known as mucopolysaccharides due to their presence in mucosa. Chemically these GAG molecules are long unbranched heteropolysaccharides composed of repeating disaccharide monomers.

The components of this disaccharide unit are:

1. *Amino sugar*: *N*-acetylgalactosamine (GalNAc) or *N*-acetylglucosamine (GlcNAc);
2. *Uronic acid*: glucuronate (GlcA) or iduronate (IdoA).

Sulfate and hydroxyl groups are present, imparting strong negative charge and extended conformation. These properties make their solution very viscous. GAGs may covalently bind with proteins to form proteoglycan. Carbohydrates may contribute >95% of proteoglycan's weight. Proteoglycans act as excellent lubricants and shock absorbers due to their water retention, great volume, and strength. There are five physiologically important GAG and include hyaluronic acid, dermatan sulfate, chondroitin sulfate, heparin, heparan sulfate, and keratan, which can be affected in mucopolysaccharidoses. These GAGs are present in bone, cartilage, tendons, cornea, skin, joints, and connective tissue as these body parts are affected by deposition.

Mucopolysaccharidoses are a group of inherited lysosomal enzyme deficiencies, which impair the catabolism of glycosaminoglycans.

Sweet Biochemistry: Remembering Structures, Cycles, and Pathways by Mnemonics
DOI: http://dx.doi.org/10.1016/B978-0-12-814453-4.00015-7

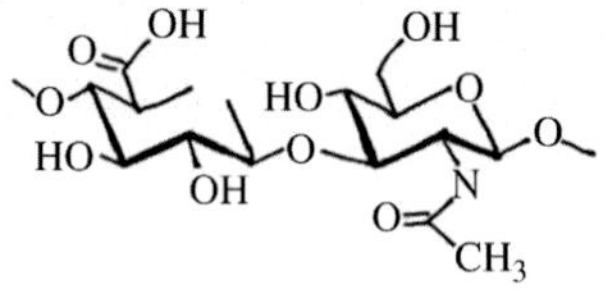

D-glucuronate (Glc A) GlcNAc

Hyaluronates:
composed of D-glucuronate(GlcA) plus GlcNAc; linkage is β(1.3)

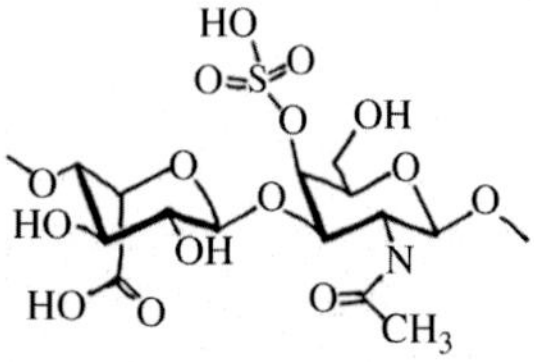

L-Iduronate (Ido A) GlcNAc-4-sulfate

Dermatan sulfates:
composed of L-iduronate (IdoA) or D-glucuronate (GlcA) plus GalNAc-4-sulfate; GlcA and IdoA sulfated; linkage is β(1.3) if GlcA, α(1,3) if IdoA

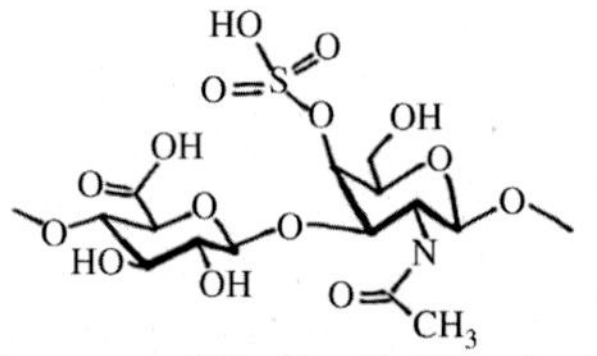

D-glucuronate (Glc A) GalNAc-4-sulfate

Chondroitin 4- and 6-sulfates:
composed of D-glucuronate (GlcA) and GalNAc-4- or 6-sulfate: linkage is β(1,3) (the figure contains GalNAc 4-sulfate)

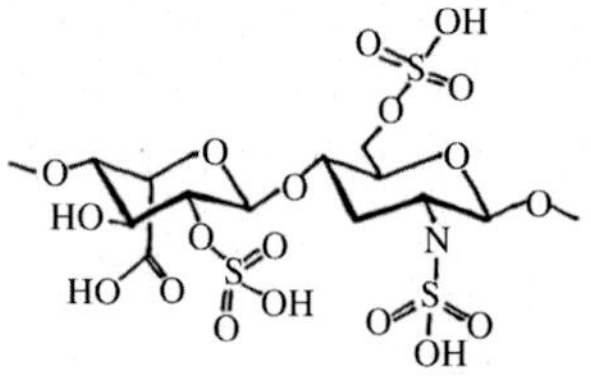

L-Iduronate-2-sulfate N-sulfo-GlcNAc-6-sulfate

Heparin and heparan sulfates:
composed of L-iduronate (IdoA: many with 2-sulfate) or D-glucuronate (GlcA: many with 2-sulfate) and *N*-sulfo-D-glucosamine-6-sulfate: linkage is α(1,4) if IdoA, β(1,4) if GlcA: heparans have less overall sulfate than heparins

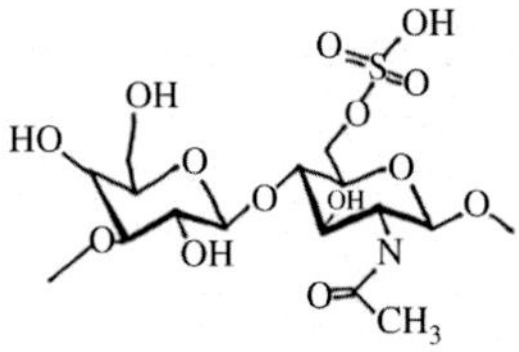

D-galactose GlcNAc-6-sulfate

Keratan sulfates:
composed of galactose plus GlsNAc-6-sulfate; linkage is β(1,4)

MUCOPOLYSACCHARIDOSES SUMMARY (TABLE 15.1)

Table 15.1 Mucopolysaccharidoses Summary

Type	Inheritance	Deficient Enzyme	GAG Accumulated
MPS 1 (Hurler, Hurler-Scheie, Scheie syndromes)	Autosomal recessive	αL-iduronidase	Dermatan sulfate, Heparan sulfate
MPSII (Hunter syndrome)	X-Linked	Iduronate-2-sulfatase	Dermatan sulfate Heparan sulfate
MPSIII (Sanfilippo syndrome)	Autosomal recessive		Heparan sulfate
Type A		Heparan sulfatase	
Type B		*N*-acetylglucosaminidase	
Type C		Acetyl-CoA glucosaminide acetyl transferase	
Type D		Acetyl glucosamine-6-sulfatase	
MPSIV (Morquio syndrome)	Autosomal	Galactose-6-sulfatase	Dermatan sulfate
Type A	recessive	Beta-galactosidase	Chondroitin sulfate
Type B			
MPSVI (Maroteaux-Lamy syndrome)	Autosomal recessive	Arylsulfatase B	Dermatan sulfate Chondroitin sulfate
MPSVII (Sly syndrome)	Autosomal recessive	Beta glucuronidase	Dermatan/heparan/ chondroitin sulfate
MPSIX	Autosomal recessive	Hyaluronidase	Hyaluron

REACTIONS ON ASKING FOR ID

Figure 15.1 Hurler, Hunter, and Scheie syndromes.

Hurler, Hunter Sir, and Scheie were asked to show their identification cards (ID—iduronidase). All three did not having their ID. Hurler said innocently that he had no idea where his ID was. Hunter Sir could not hear the question properly due to deafness. Scheie said very smartly: "IDs can be fake so why worry so much about them."

Notice the following: Hurler and Scheie are wearing black goggles due to corneal opacity. Hunter has an X mark on his shirt because of X-linked inheritance. He is called Sir to indicate sulfatase. Hurler and Hunter are both mentally retarded. Skeletal deformity is present in all three syndromes (Fig. 15.1).

HURLER DISEASE: MPS1H

Figure 15.2 Hurler syndrome.

- The enzyme lysosomal alpha-L-iduronidase, which degrades mucopolysaccharides, is deficient in Hurler syndrome (Fig. 15.2)
- It is the most severe type of MPS I.
- Deposition of heparan sulfate and dermatan sulfate takes place
- Affected individuals are normal at birth but exhibit the following features
  - Severe mental retardation;
  - Progressive deterioration;
  - Hepatosplenomegaly;
  - Corneal clouding and retinal degeneration;
  - Deafness and enlarged tongue;
  - Dwarfism with a hunched back;
  - Nerve compression;
  - Restricted joint movements;
  - Cardiac failure due to infiltration of the heart;

SCHEIE SYNDROME: MPS1S

Figure 15.3 Scheie syndrome.

- The enzyme lysosomal alpha-L-iduronidase, which degrades mucopolysaccharides, is deficient in Scheie syndrome.
- It is the least severe form of Hurler syndrome.
- Deposition of heparan sulfate and dermatan sulfate takes place.
- The most common features of Scheie syndrome include:
  - Normal intelligence;
  - Less progressive deterioration and approximate normal lifespan;
  - Corneal clouding;
  - Restricted joint movements;
  - Valvular heart disease (Fig. 15.3).

HUNTER SYNDROME: MPS II

Figure 15.4 Hunter syndrome.

Hunter syndrome is caused by deficiency or absence of iduronate-2-sulfatase (I2S) in lysosomes (Fig. 15.4).

It is the only MPS which is X-linked.

Excesses of heparan sulfate and dermatan sulfate are accummulated in organs including:

- *Lungs*: limited lung capacity;
- *Heart valves*: decreased cardiac function;
- *Liver, spleen*: hepatosplenomegaly;
- *Joints*: joint stiffness;
- *Brain*: mental retardation.

Other common features include:

- Abdominal hernias;
- Repeated ear and respiratory tract infections, deafness;
- Coarse facial features.

SANFILIPPO SYNDROME: MPS III

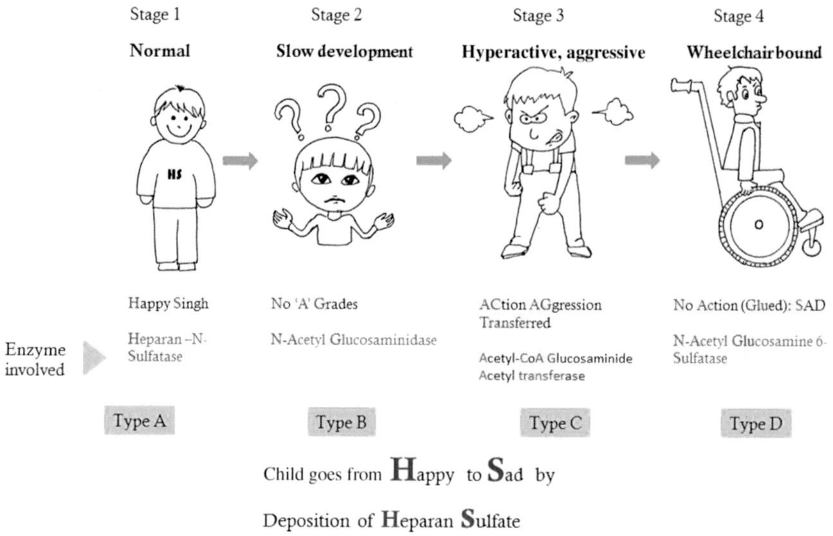

Figure 15.5 Sanfilippo syndrome.

In Fig. 15.5 you can see the stages of a Sanfilippo child. The first stage is a normal baby, the second stage shows slow development in getting no "A" grades, the third stage is a hyperactive, angry child, and the fourth stage is a wheelchair-bound child. Different types of Sanfilippo syndrome along with the enzyme involved are also illustrated, correlated with mnemonics. Sanfilippo syndrome is due to accumulation of heparan sulfate.

Sanfilipo syndrome is caused by deficiency or absence of the enzyme involved in degradation of heparan sulfate in lysosomes. It is an autosomal recessive disease. Heparan sulfate is a glycosaminoglycan present in the extracellular matrix and cell surface. Four enzyme defects have been discovered, namely:

- Heparan *N*-sulfatase;
- *N*-acetyl-glucosaminidase;
- Acetyl-CoA glucosaminide acetyl transferase;
- *N*-acetyl glucosamine-6-sulfatase.

Patients with Sanfilippo syndrome present with mild facial dysmorphism, stiff joints, hirsutism, and coarse hair as in other MPS patients. However, in this syndrome the progression follows a pattern of four stages. Initially the child appears to be normal, but in a few months to years development slows. Progressive motor disease, severe dementia, and poor performance in studies are common complaints in the second stage. In the third stage behavioral disturbances appear with hyperactivity, aggressiveness, pica, and disturbances in sleep. In the final stage the child becomes unable to walk, and is wheelchair-bound, along with seizures and swallowing difficulties. These patients usually live up to their early twenties or may die younger.

MORQUIO SYNDROME: MPS IV

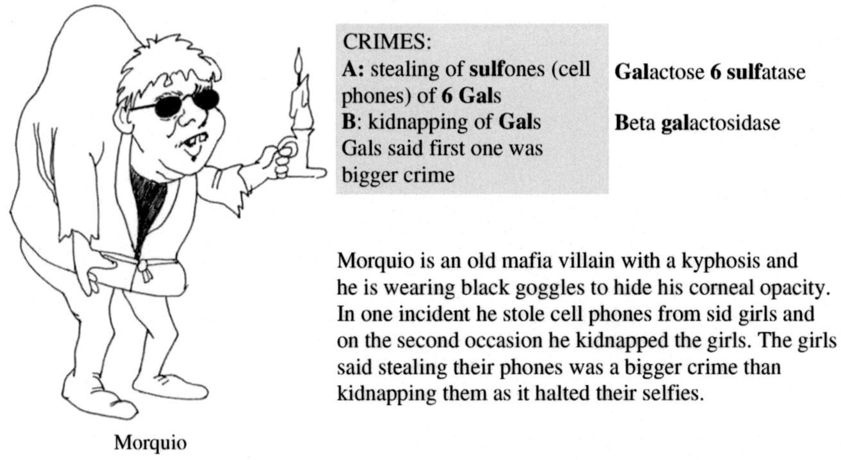

CRIMES:
A: stealing of **sulf**ones (cell phones) of **6 Gal**s
B: kidnapping of **Gal**s
Gals said first one was bigger crime
Galactose **6 sulf**atase
Beta **gal**actosidase
Morquio is an old mafia villain with a kyphosis and he is wearing black goggles to hide his corneal opacity. In one incident he stole cell phones from sid girls and on the second occasion he kidnapped the girls. The girls said stealing their phones was a bigger crime than kidnapping them as it halted their selfies.
Morquio

Figure 15.6 Morquio syndrome.

In Morquio syndrome (Fig. 15.6) keratan sulfate (KS) is piled up in tissues because of deficiency of either *N*-acetyl-galactosamine-6-sulfate sulfatase in Morquio syndrome type IVA or β-galactosidase in Morquio syndrome type IVB. Catabolism of chondroitin 6-sulfate is also affected. KS is abundantly present in the cornea and cartilage, hence these tissues are majorly affected. Type IVA is a more severe form than IVB. Affected individuals present with mental retardation, skeletal deformity (e.g., scoliosis, kyphosis, and rib flaring), epiphyseal dysplasia, corneal opacity, and odontoid hypoplasia.

MAROTEAUX-LAMY SYNDROME: MPS VI

Figure 15.7 Maroteaux-Lamy syndrome.

Maroteaux-Lamy is the deficiency of N-acetylgalactosamine 4-sulfatase. GAG deposited is dermantan sulfate Figs. 15.7 and 15.8.

Characteristic features include:

- Dysostosis multiplcx, short stature
- Corneal clouding, deafness, dural thickening, and pain caused by nerve compression developmental dclay
- Heart defects

SLY SYNDROME: MPS VII

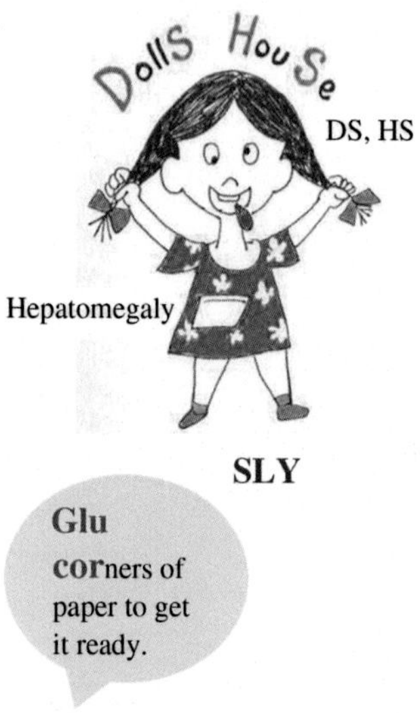

Sly is a cute little mentally retarded girl requesting a **DollS HouSe** (DS, HS is deposited in Sly syndrome). Her mother tells her to make her own dolls house by gluing corners of paper.

Figure 15.8 Sly syndrome.

Sly syndrome is a lysosomal storage disease caused by deficient beta glucuronidase leading to accumulation of dermantan sulfate, heparan sulfate and chondroitin sulfate.

Major clinical manifestations include the following:

- Dysmorphic facial features, such as coarse face, macrocephaly, frontal prominence are present
- Corneal clouding or opacity
- Hepatosplenomegaly
- Skeletal features–dysostosis multiplex, kyphosis
- Hernias and vascular complications arise due to connective tissue involvement
- Mental retardation and development delay
- Wide spectrum of presentations: hydrops fetalis in severe form and a neonatal form with jaundice is also present

CHAPTER 16

Prostaglandin Synthesis

TRADITIONAL PROSTAGLANDIN SYNTHESIS RECAP

Prostaglandins (PGs) are physiologically and pharmacologically active, lipid compounds resembling hormones which act through G-protein-linked receptors (Figs. 16.1 and 16.2). PGs are found in almost all nucleated cells of the body but the name prostaglandin was coined due to their discovery in the prostate gland. They elicit their biological hormone-like actions locally as they are short-lived. The effects of PGs vary in different cells because of binding to different receptors. The principal actions of PGs include inflammation stimulation, regulation of blood flow, sleep cycle induction, affecting membrane transport, and modulating synaptic transmission in the nervous system.

Prostaglandins are derived from essential fatty acids of membrane. Chemically, PGs are C20 saturated fatty acid derivatives with a five-carbon ring. PGs belong to the eicosanoids family (Greek eikosi means 20), which arise from arachidonic acid. Other members of the eicosanoids are thromboxanes (TX), leukotrienes (LT), and lipoxins (LX). Arachidonic acid (AA) can be a substrate for two enzymes: cyclooxygenase (COX) and lipooxygenase (LOX). When AA is a substrate for COX, PG2, TX2 series are synthesized, while on entering the LOX pathway, LT4 and LX4 series are formed. According to the ring present in a PG, it is given a letter which is followed by a number indicating the number of double bonds, for example, PGE1, PGE2, PGI2, PGF2α. Alpha denotes the OH projection on carbon 9 in PGs.

Arachidonic acid, a 20:4 fatty acid derived from linoleate essential fatty acid, undergoes sequential oxidations and isomerizations to form PG. COX enzyme catalyzing first committed step is of two

Sweet Biochemistry: Remembering Structures, Cycles, and Pathways by Mnemonics
DOI: http://dx.doi.org/10.1016/B978-0-12-814453-4.00016-9

types COX-1 and COX-2. COX-2 is involved in inflammation and growth, while COX-1 is a product of the housekeeping gene. These enzymes are mainly active in the blood vessels, kidneys, and stomach.

In PG synthesis, arachidonic acid is converted to PGG2 by action of COX. PGG2 is oxidized to PGH2, which can be further isomerized to PGD2, PGE2, and PGF2α.

Prostaglandin E2 (PGE2) is generated from the action of prostaglandin E synthases on prostaglandin H2 (prostaglandin H2, PGH2).

Alternatively, PGH2 can be a substrate for thromboxane synthase to yield TXA2 and TXB2. PGH2 can also form prostacyclins in the presence of prostacyclin synthase.

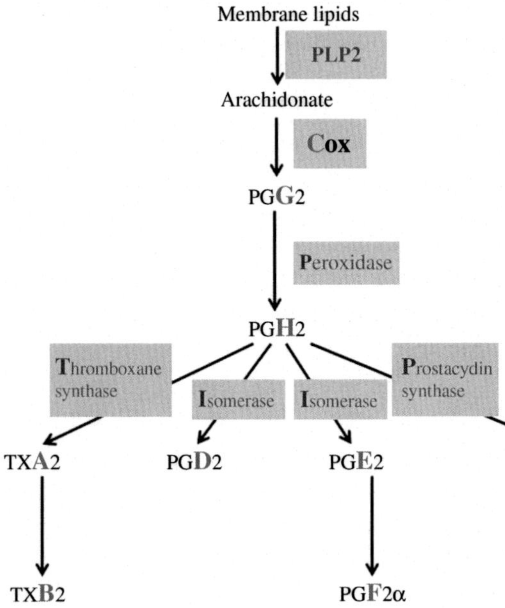

Figure 16.1 Prostaglandin synthesis.

PROSTAGLANDIN SYNTHESIS MNEMONIC

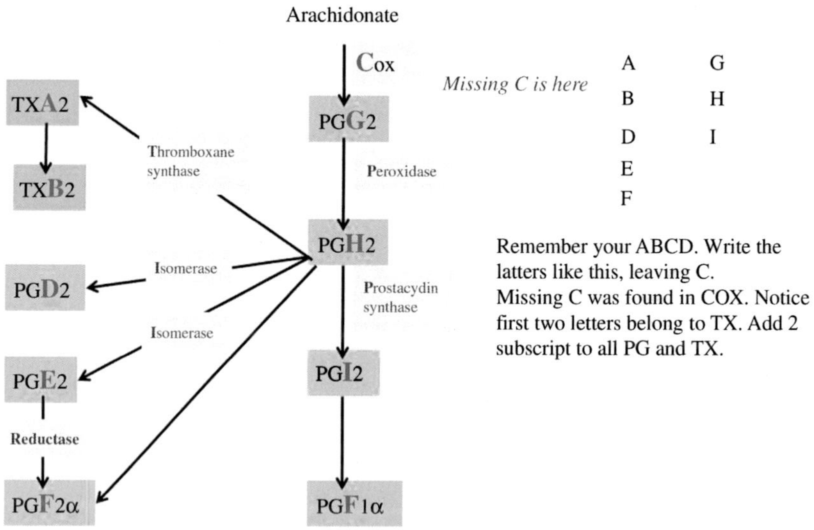

Figure 16.2 Prostaglandin synthesis mnemonic.

CHAPTER 17

Purine Structures

TRADITIONAL PURINE STRUCTURE RECAP

A purine is an aromatic heterocycle composed of carbon and nitrogen (Fig. 17.1). Purines include adenine and guanine, which participate in DNA and RNA formation. Purines are also constituents of other important biomolecules, such as ATP, GTP, cyclic AMP, NADH, and coenzyme A. Purines have an NH_2 group and oxo groups which exhibit keto-enol and amine-imine tautomerism, although amino and oxo forms predominate in physiological conditions.

Basic purine has nine atoms in its structure. Purine has two cycles: a six-membered pyrimidine ring and a five-membered imidazole ring fused together. Four nitrogen atoms are present at the 1, 3, 7, and 9 positions. The numbering of purine starts with the first nitrogen of the six-membered ring and then proceeds in an anticlockwise direction. The imidazole ring is numbered clockwise. Other important purines include hypoxanthine, xanthine, theobromine, caffeine, uric acid, and isoguanine. Purine bases connect with carbon-1' of pentoses through the ninth nitrogen atom to form nucleosides.

The sources of atoms of purine bases are:

- Nitrogen 1—amino group of aspartate;
- Carbon 2—formyl THFA;
- Nitrogen 3—amide *N* of glutamine;
- Carbon 4, carbon 5, nitrogen 7—glycine;
- Carbon 6—respiratory CO_2;
- Carbon 8—methylene THFA;
- Nitrogen 9—amide *N* of glutamine.

Chemically:

- Adenine = 6-amino purine;
- Guanine = 2-amino, 6-oxypurine;
- Hypoxanthine = 6-oxy purine;
- Xanthine = 2,6-dioxy purine;
- Uric acid = 2,6,8-tri-oxopurine.

Sweet Biochemistry: Remembering Structures, Cycles, and Pathways by Mnemonics
DOI: http://dx.doi.org/10.1016/B978-0-12-814453-4.00017-0

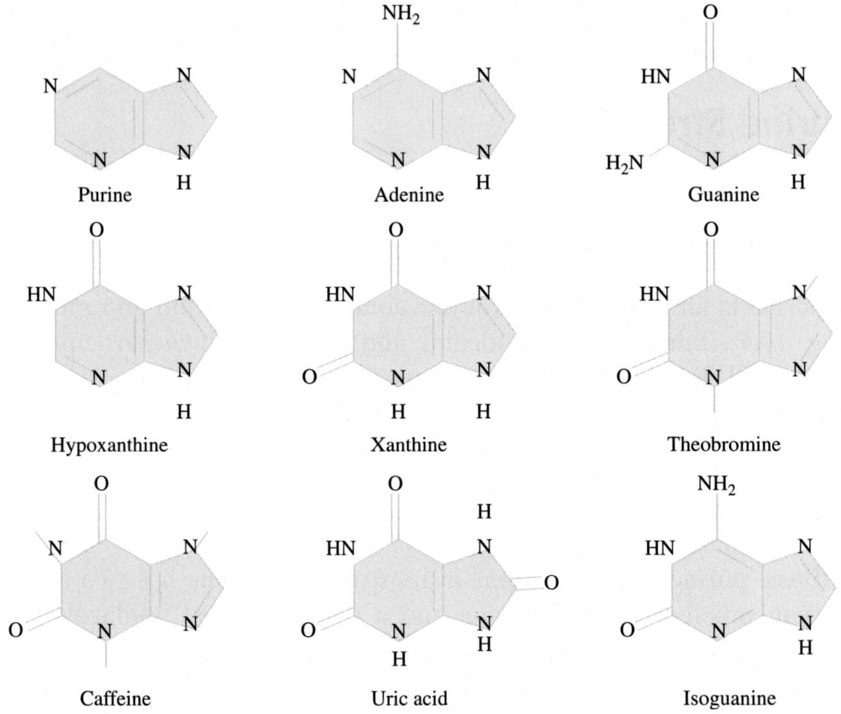

Figure 17.1 Structures of various purines.

PICTURE MNEMONIC FOR SOURCES OF ATOMS OF PURINE

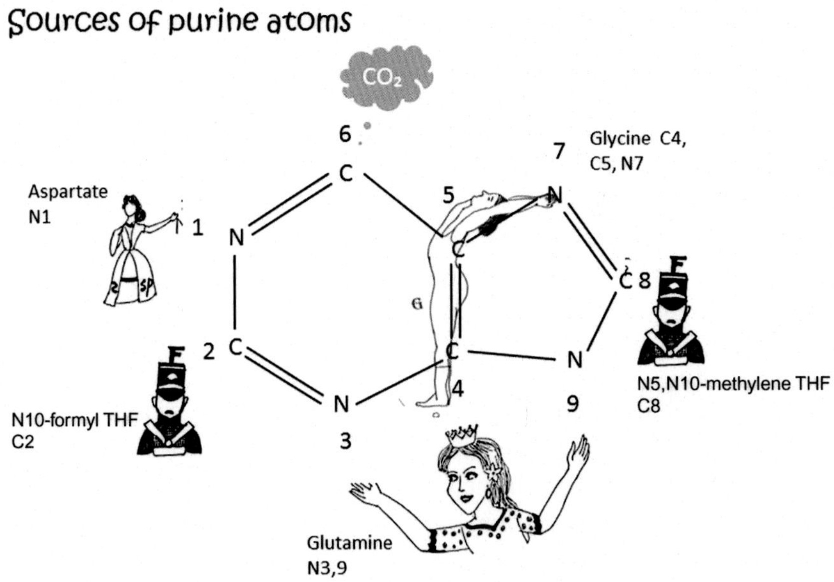

Figure 17.2 Picture mnemonic for sources of atoms of purine.

Different sources of atoms of purine are depicted in Fig. 17.2. For nitrogen 1, a lady wearing a dress with "Asp" written on it is drawn. Carbon 2 and carbon 8 are donated from formyl THFA and methylene THFA soldiers; notice their caps with a big "F" for folate. Nitrogen 3 and 9 are gifted by Princess Glutamine. Carbons 4, 5 and Nitrogen 7 are embedding a glycine in the structure. Lastly, carbon 6 is taken from clouds of carbon dioxide.

CHAPTER 18

Purine de novo Synthesis

TRADITIONAL PURINE DE NOVO SYNTHESIS RECAP

- Purines are mainly synthesized by the liver through the cytoplasmic de novo synthesis pathway (Figs. 18.1–18.4). De novo synthesis means that the rings are made by compiling atoms from their sources. Purine rings are synthesized on a platform of ribose-5-phosphate to yield nucleotides. The de novo synthesis pathway takes place in 10 steps. The enzymes of this pathway form a multienzyme complex in eukaryotes for better efficiency. Phosphoribosyl pyrophosphate (PRPP) provides ribose-5-phosphate as the starting material.
- In the first reaction, which is also the committed step, ammonia (N9 of final purine) released from glutamine displaces pyrophosphate to produce *5-phosphoribosyl-1-amine*. The enzyme governing this step is *glutamine phosphoribosyl amidotransferase*.
- The following reactions share a similar mechanism in which a carbon-bound oxygen atom is activated by phosphorylation and then the phosphoryl group is displaced by ammonia or an amine group acting as a nucleophile.
- The glycine molecule is almost completely consumed (C4, C5, N7) in the growing ring by joining the amino group of phosphoribosylamine.
- In the next step, a formyl group (C8) is transferred to an amino group of the glycine component by *N*-10-formyltetrahydrofolate.
- Before closure of the five-membered imidazole ring, an amidine group is formed by ammonia released from glutamine. This adds a third nitrogen (N3) to the purine ring.
- The five-membered imidazole ring is closed utilizing an ATP.
- Carbon dioxide in the form of bicarbonate adds C6 to the imidazole ring.
- The imidazole carboxylate is phosphorylated, followed by displacement of the phosphate by the amino group of aspartate.
- Fumarate is released, leaving behind an amino group.
- A second formyl group (C2) is added from *N*-10-formyl THFA and second cyclization yields inosinate (IMP).

Sweet Biochemistry: Remembering Structures, Cycles, and Pathways by Mnemonics
DOI: http://dx.doi.org/10.1016/B978-0-12-814453-4.00018-2

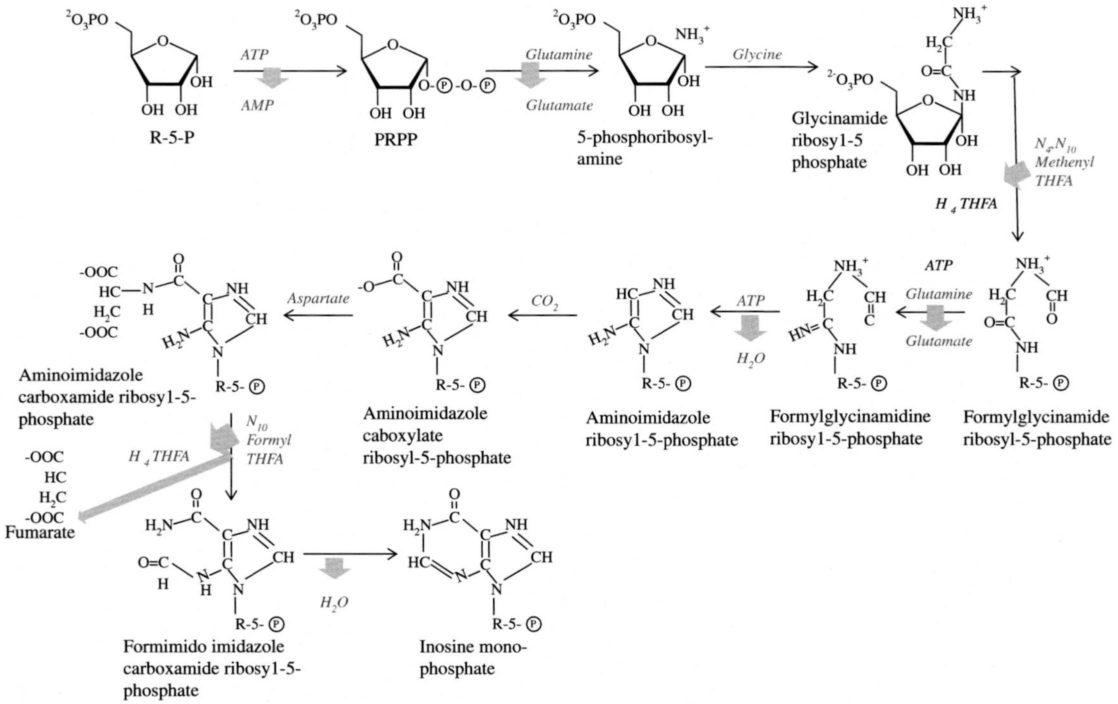

Figure 18.1 Purine de novo synthesis with structures.

Enzymes of Purine de novo synthesis recap

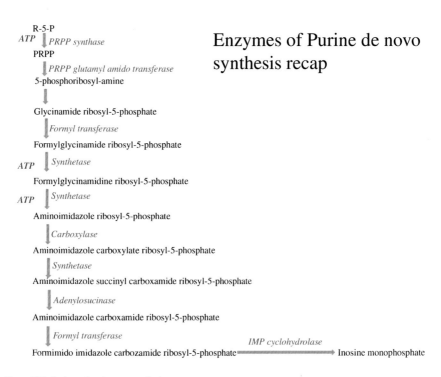

R-5-P
↓ *PRPP synthase* (ATP)
PRPP
↓ *PRPP glutamyl amido transferase*
5-phosphoribosyl-amine
↓
Glycinamide ribosyl-5-phosphate
↓ *Formyl transferase*
Formylglycinamide ribosyl-5-phosphate
↓ *Synthetase* (ATP)
Formylglycinamidine ribosyl-5-phosphate
↓ *Synthetase* (ATP)
Aminoimidazole ribosyl-5-phosphate
↓ *Carboxylase*
Aminoimidazole carboxylate ribosyl-5-phosphate
↓ *Synthetase*
Aminoimidazole succinyl carboxamide ribosyl-5-phosphate
↓ *Adenylosucinase*
Aminoimidazole carboxamide ribosyl-5-phosphate
↓ *Formyl transferase*
Formimido imidazole carbozamide ribosyl-5-phosphate → *IMP cyclohydrolase* → Inosine monophosphate

Figure 18.2 Basic purine de novo synthesis.

TRICK TO LEARN PURINE DE NOVO SYNTHESIS

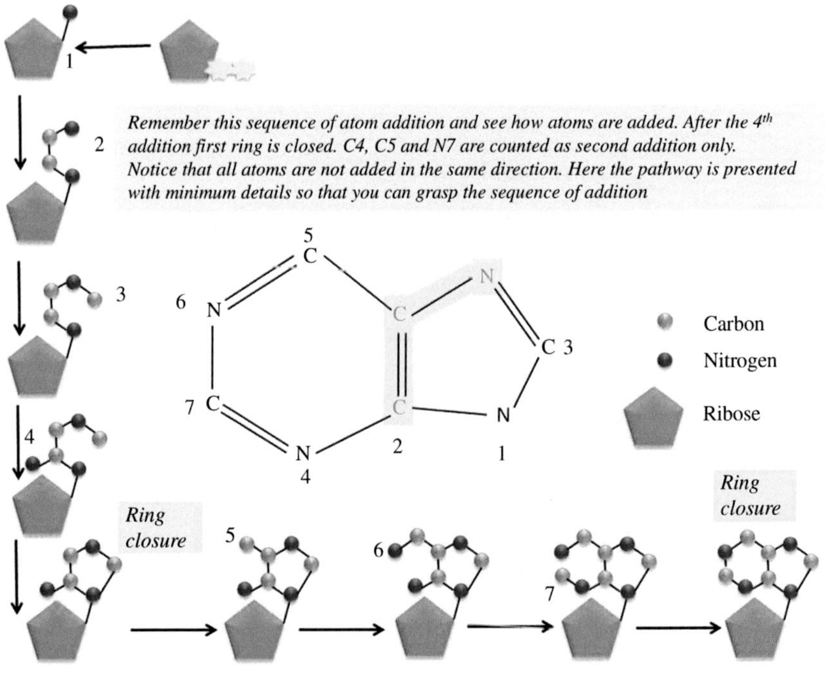

1
2
Remember this sequence of atom addition and see how atoms are added. After the 4th addition first ring is closed. C4, C5 and N7 are counted as second addition only. Notice that all atoms are not added in the same direction. Here the pathway is presented with minimum details so that you can grasp the sequence of addition
3
4
5
C
6
N
7
C
N
4
C
C
2
N
C 3
N
1
Carbon
Nitrogen
Ribose
Ring closure
5
6
7
Ring closure

Figure 18.3 Sequence of addition of purine atoms—an easy way to learn the pathway.

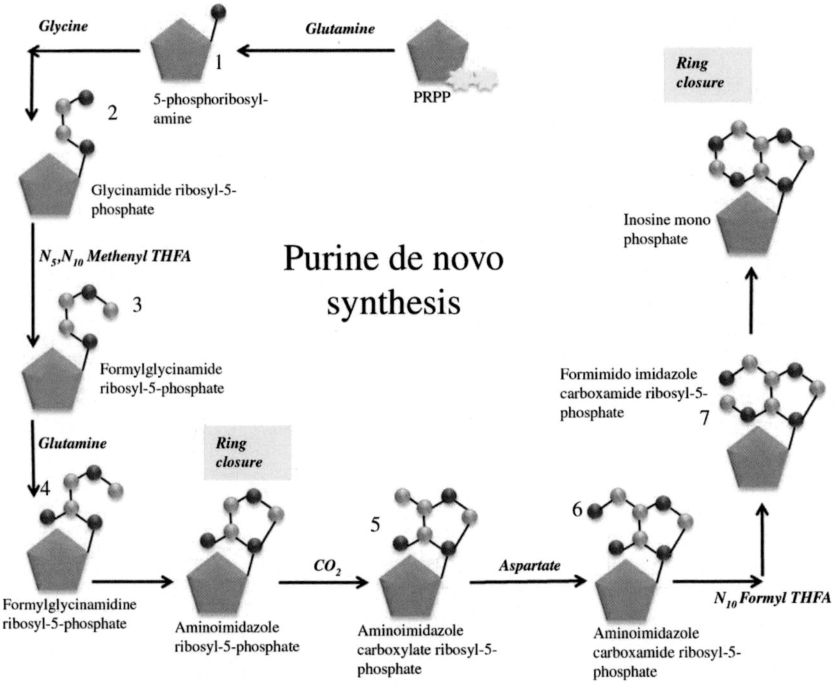

Figure 18.4 Correlating the addition of purine atoms with substrates.

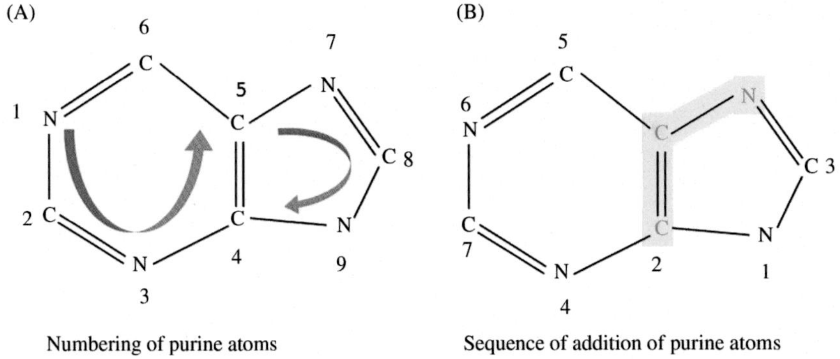

Figure 18.5 Differentiating the numbering of purine atoms from the sequence of addition of a purine atom.

In Fig. 18.5, the reader must not confuse the numbering of purine atoms with the sequence of atom additions. Numbering of the purine ring is shown in (A), where the pyrimidine part is numbered anticlockwise and the imidazole ring is numbered clockwise.

The sequence of addition of purine atoms is shown in (B), which is not in a linear direction. The sequence of addition of atoms when used with sources of atoms can indicate the reactions. This simplifies this intricate pathway.

CHAPTER 19

Pyrimidine Structure

PYRIMIDINE STRUCTURE RECAP

The pyrimidine ring is an aromatic heterocycle of two nitrogen and four carbon atoms. The numbering of atoms is done in a clockwise direction. Nitrogen atoms are present at positions 1 and 3. The sources of carbon 2 and nitrogen 3 are carbamoyl phosphate, while the rest of the ring is derived from aspartate. Only nitrogen 1 of the pyrimidine ring forms a glycosidic linkage with C-1′ of ribose sugar.

Uracil, cytosine, and thymine are the principal pyrimidines which constitute uridine, cytidine, and thymidine ribonucleosides and the corresponding deoxynucleosides. Cytosine and thymine are the building blocks of DNA, while cytosine and uracil are found in RNA. The pyrimidine ring has a planar structure, this helps in stacking interactions with purine bases. Pyrimidine pairs with complementary purine bases by hydrogen bonding, for example, thymine with adenine, and cytosine with guanine. Thymine is 5-methyluracil—it is very interesting that uracil is present in RNA, while its methylated form, thymine, is seen in DNA. The explanation given for this is that cytosine can deaminate to form uracil, which is repaired by the DNA repair system. If uracil was already present in DNA it would be very difficult for the repair system to identify the component uracil from deaminated cytosine. Hence, thymine replaced uracil in DNA.

Pyrimidines form not only building blocks of DNA and RNA but they also serve important functions like polysaccharide and phospholipid synthesis, glucuronidation in detoxification, and glycosylation of proteins and lipids. The structures of important pyrimidines are shown in Figs. 19.1 and 19.2.

The pyrimidine ring can be compared with a clock. The numbering of pyrimidine is done in a clockwise manner and the sugar is attached to pyrimidine at N1, resembling a pendulum hanging from a clock (Fig. 19.3).

Sweet Biochemistry: Remembering Structures, Cycles, and Pathways by Mnemonics
DOI: http://dx.doi.org/10.1016/B978-0-12-814453-4.00019-4

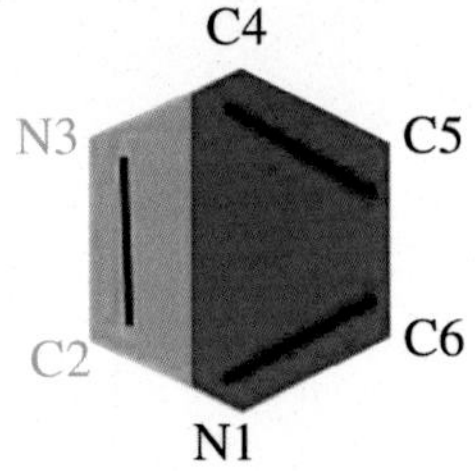

Figure 19.1 Pyrimidine ring.

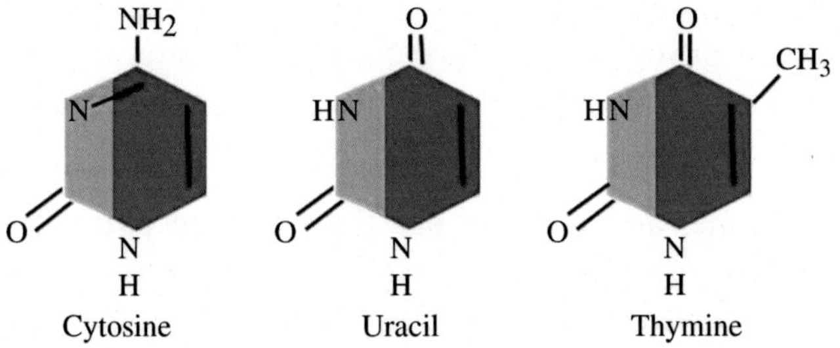

Figure 19.2 Important pyrimidine bases.

PICTURE MNEMONIC FOR SOURCES OF ATOMS OF PYRIMIDINE

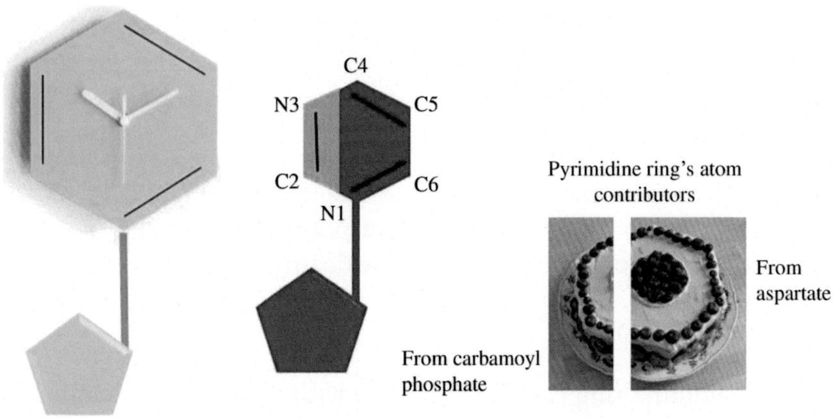

Figure 19.3 Pyrimidine with sugar compared to a wall clock with a pendulum.

CHAPTER 20

Pyrimidine de novo Synthesis

PYRIMIDINE DE NOVO SYNTHESIS RECAP

Pyrimidine synthesis takes place in cytoplasm. Pyrimidine is synthesized as a free ring and then a ribose-5-phosphate is added to yield direct nucleotides, whereas, in purine synthesis, the ring is made by attaching atoms on ribose-5-phosphate. The first three enzymes (blue-colored) and fifth and sixth enzymes (green-colored) are part of two multifunctional peptides to increase efficiency.

The important steps of pyrimidine synthesis are:

1. *Carbamoyl phosphate synthase II step*—Carbamoyl phosphate synthetase II catalyzes the reaction of bicarbonate and ammonia from glutamine in the cytoplasm to produce carbamoyl phosphate. This enzyme is different from CPS I involved in urea synthesis.
2. *Aspartate transcarbamoylase step*—The second main source of pyrimidine ring aspartate combines with carbamoyl phosphate in the presence of aspartate transcarbamoylase. This step is the committed step of the pathway as this enzyme is allosterically regulated (allosteric inhibition by CTP).
3. *Dihydro-orotase step*—Covalent bonding between N3 and C4 closes the ring, yielding dihydro-orotate. The enzyme participating is dihydro-orotase. These three enzymes are together called CAD and this is a multifunctional protein.
4. *Dihydro-orotase dehydrogenase step*—A double bond between C5 and C6 is formed by dihydro-orotate dehydrogenase utilizing NAD+ as a coenzyme. Orotic acid is formed in this reaction.
5. *Orotase phosphoribosyl transferase*—Orotic acid is converted to orotidine monophosphate (OMP) by orotate phosphoribosyl transferase. Here ribose-5-phosphate from PRPP is attached to N1 of orotic acid, releasing pyrophosphate.

Sweet Biochemistry: Remembering Structures, Cycles, and Pathways by Mnemonics
DOI: http://dx.doi.org/10.1016/B978-0-12-814453-4.00020-0

6. *Orotidylic acid decarboxylase*—Decarboxylation of OMP is catalyzed by orotidylic acid decarboxylase. Carbon 7 of the ring is removed as carbon dioxide yielding uridine monophosphate UMP (Figs. 20.1 and 20.2). The last two enzymes are also present as a multifunctional protein.

UMP can be phosphorylated to form UDP and UTP. CTP can be synthesized by adding an amino group from glutamine to UTP.

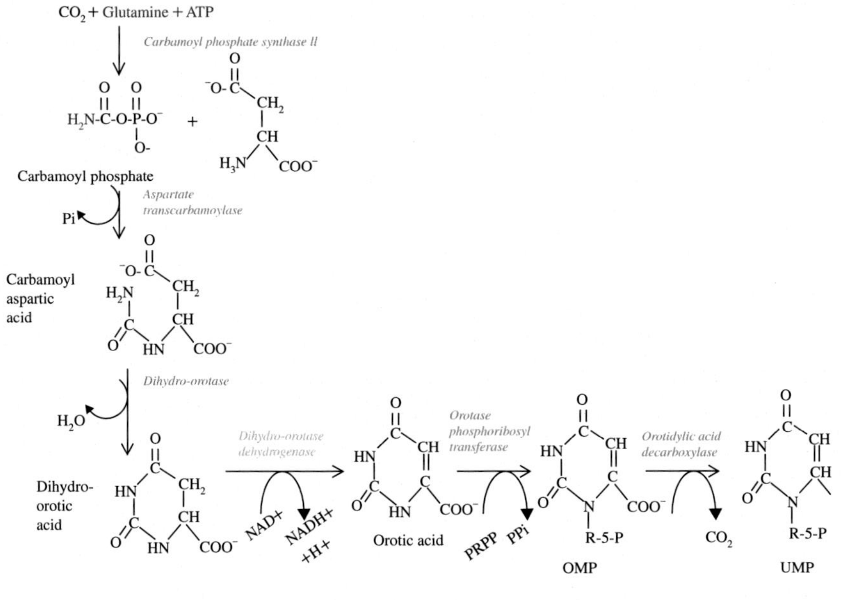

Figure 20.1 Pyrimidine de novo synthesis.

SIMPLIFIED PYRIMIDINE DE NOVO SYNTHESIS

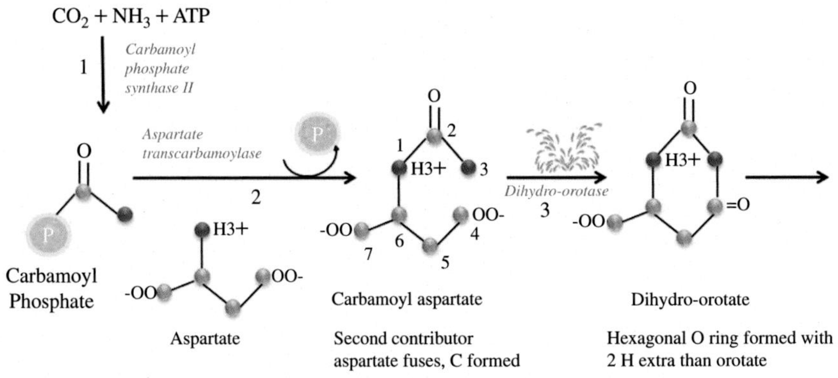

Figure 20.2 Simplified pyrimidine synthesis—1.

It is easier to remember pyrimidine synthesis by structures (Fig. 20.3). Remember the sources of ring atoms: carbamoyl phosphate and aspartate. First prepare a small molecule, i.e., carbamoyl phosphate. As the name indicates, it is generated from carbon dioxide (carb), ammonia (amo), and ATP (phosphate). Ammonia in this reaction comes from glutamine. We try to make a ring with our substrates, carbamoyl phosphate and aspartate. They join to form a C-shaped open ring, which is joined by dehydration and polished by dehydrogenation. After adding a diamond on the ring and cutting the extra carbon projections, the ring is ready.

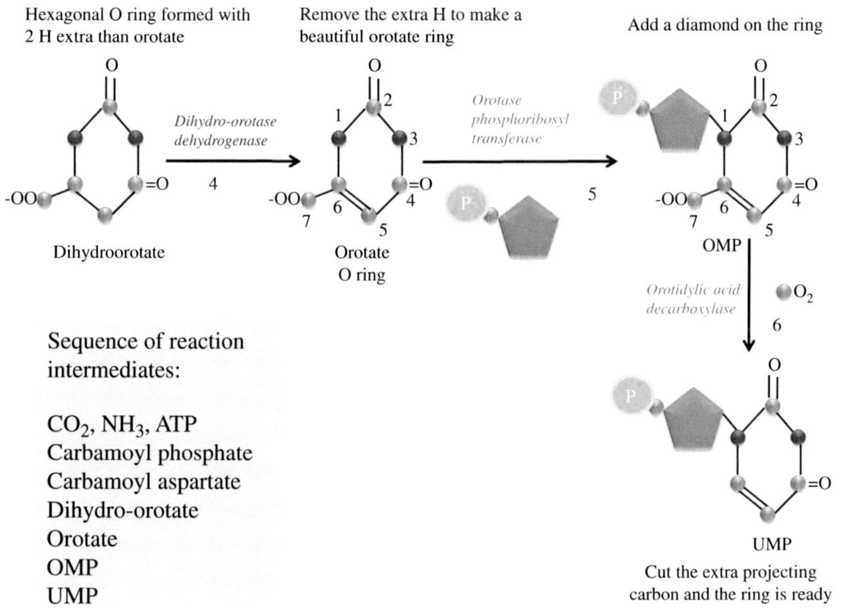

Figure 20.3 Simplified pyrimidine synthesis—2.

EXERCISES

After you read any of the chapters in this book, try to complete these exercises and enjoy…

CHAPTER 1

Write the glycolysis intermediates in the front of the poem.

Give
Give 6 peg
For my 6 phrn (friend)
One more peg
For me then
Cut in two
BPG is new
He is very strong
One ATP is gone
Three two one
Add PEPPER in fun
Second ATP runs
And the party ends.

Write the missing letters of the mnemonic and then write the reaction names in the box below.

See it and remember it

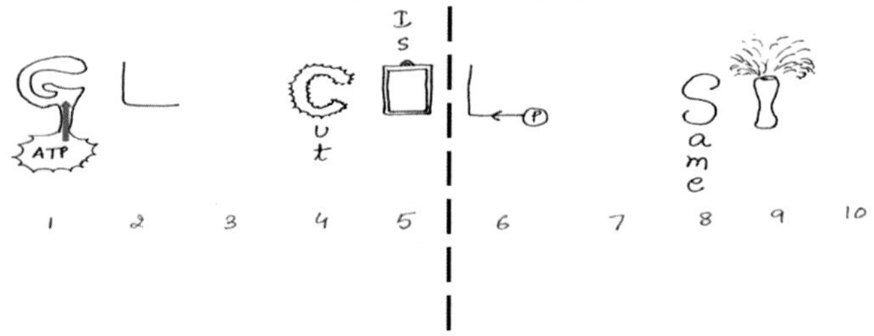

CHAPTER 2

Write the story below the pictures and see how much you remember.

Krebs Cycle

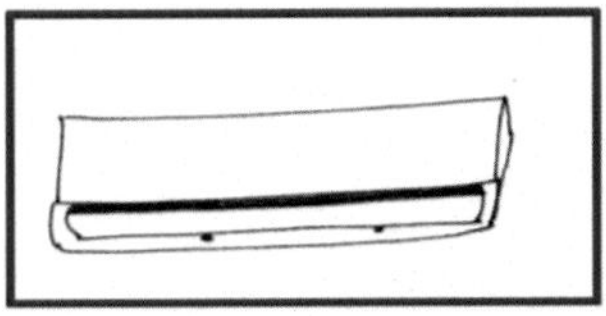

Write the name of the intermediates of the Krebs cycle matching with the story.

Write the Krebs cycle mnemonic.

Write krebs cycle with numbering starting from K. write S and C slightly higher than other letters

Water entry/ addition

CO_2 release/ Decarboxylation step

NAD+ dehydrogenation steps

FAD dehydrogenation steps

CoA-SH release

CHAPTER 3

See the story of ETC from Chapter 3, Electron Transport Chain and write below the ETC components correlating with them.

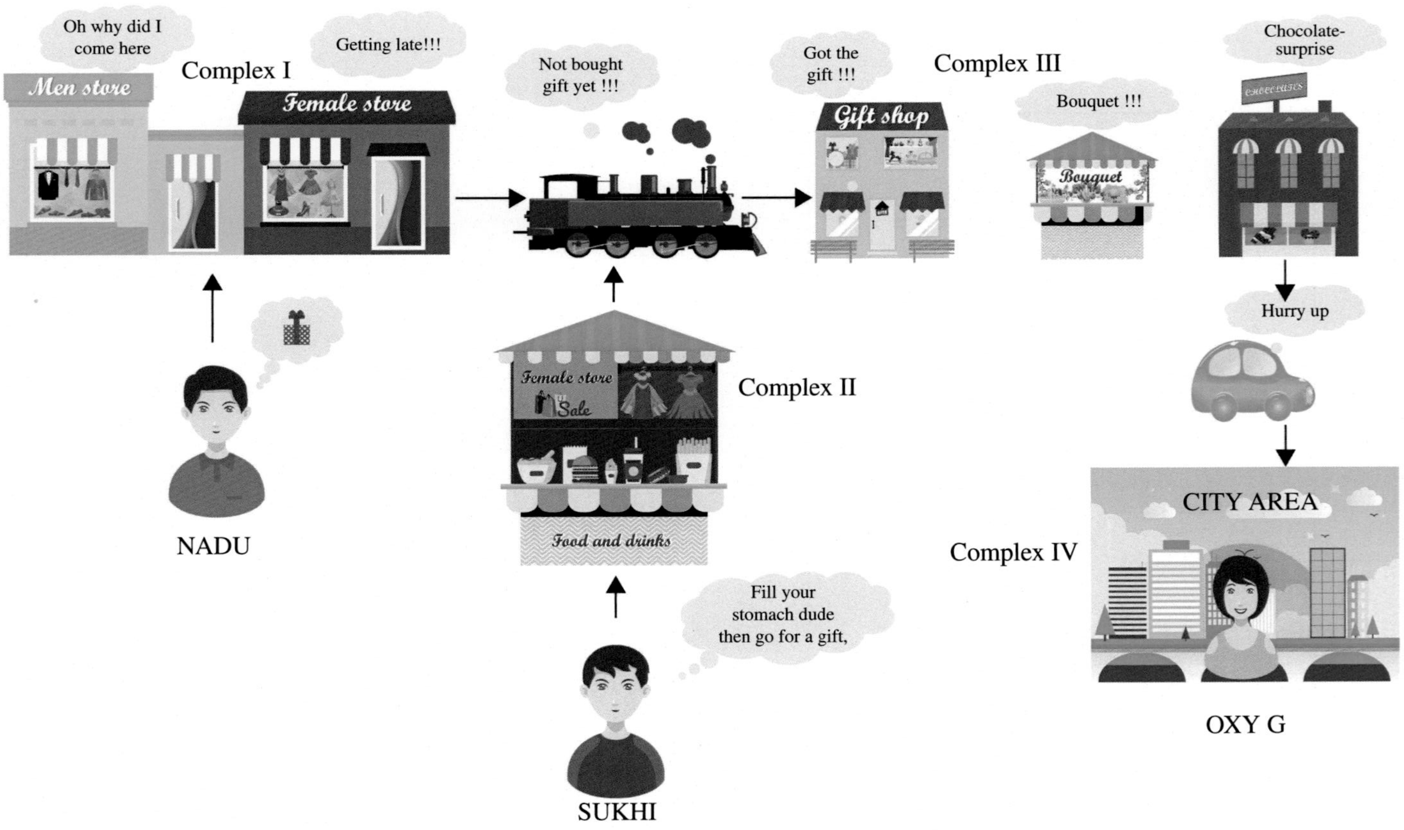
Oh why did I come here
Complex I
Men store
Female store
Getting late!!!
Not bought gift yet !!!
Got the gift !!!
Complex III
Gift shop
Bouquet !!!
Bouquet
Chocolate-surprise
Hurry up
NADU
Female store
Sale
Complex II
Food and drinks
Fill your stomach dude then go for a gift,
SUKHI
Complex IV
CITY AREA
OXY G

NADU →
For MeN store →
Female Store →
Train →
City Bouquet shop →
City Chocholate shop →
Car →
City AreA3 →
Oxy G →
Sukhi →
Food And Drinks →

CHAPTER 4

Complete the diagram.

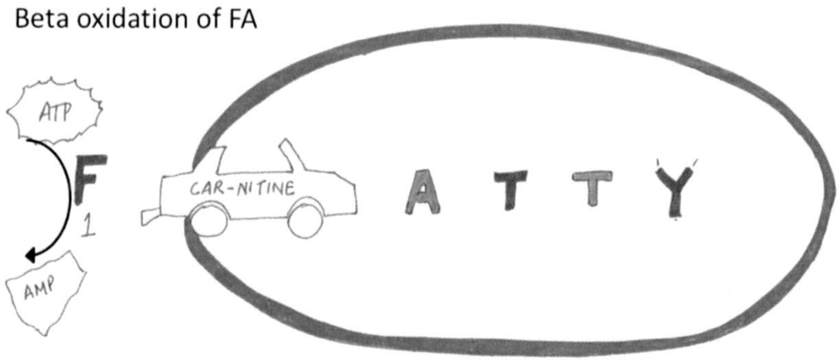

CHAPTER 5

Complete the diagram.

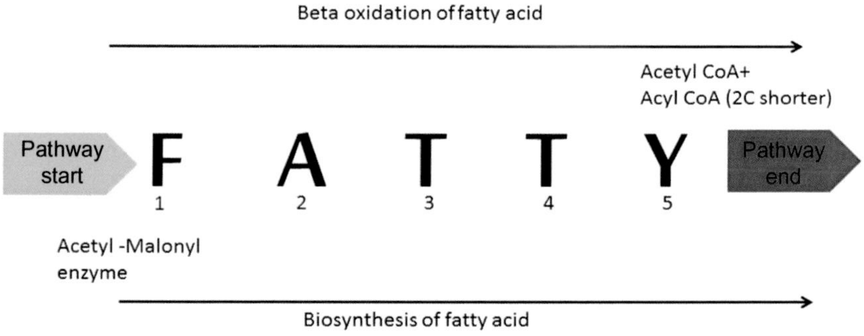

CHAPTER 6

See the Queen Bee's house and draw the actual cholesterol structure.

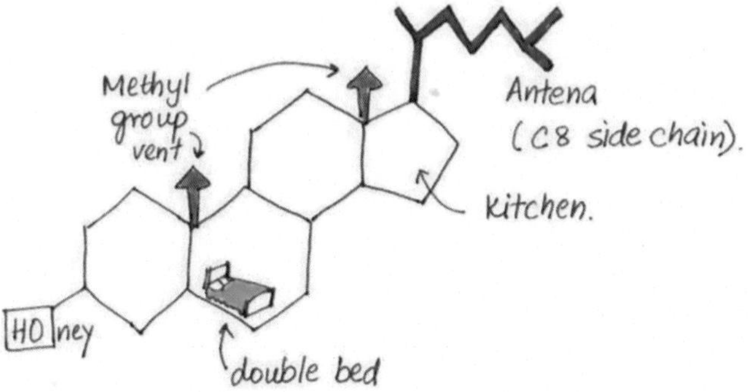

CHAPTER 7

Write in the table below the steps of cholesterol synthesis which correlate with the story.

What Bee Does	What this Correlates to in the Cholesterol Pathway
2C	
4C	
6C	
5C	
10C	
15C	
30C	
Ouspicious (auspicious) board at the entrance of house	
She rearranges the walls	
She discards the extra hangings below the floor	
Next she moved her bed	
She corrected her TV antenna	

CHAPTER 8

Fill the instructions given to the dancers and the related intermediate name.

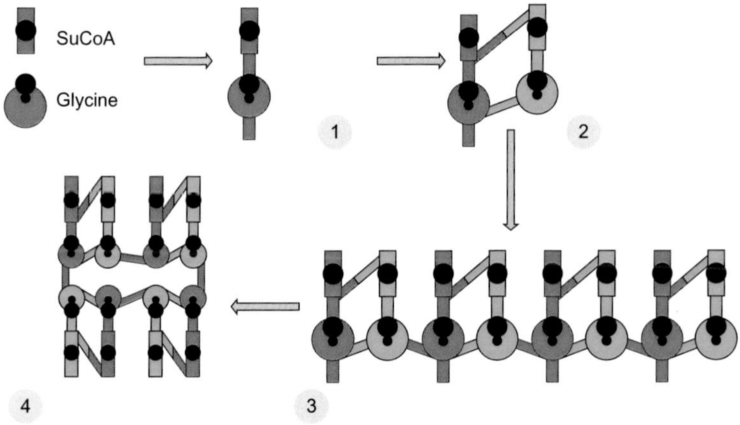

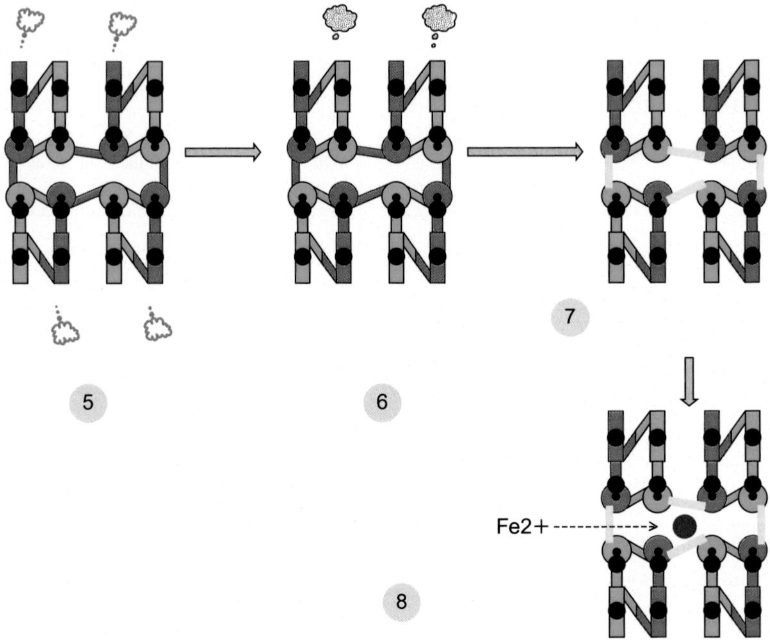

CHAPTER 9

Write the corresponding porphyria according to the diagram.

PORPHYRIAS summary

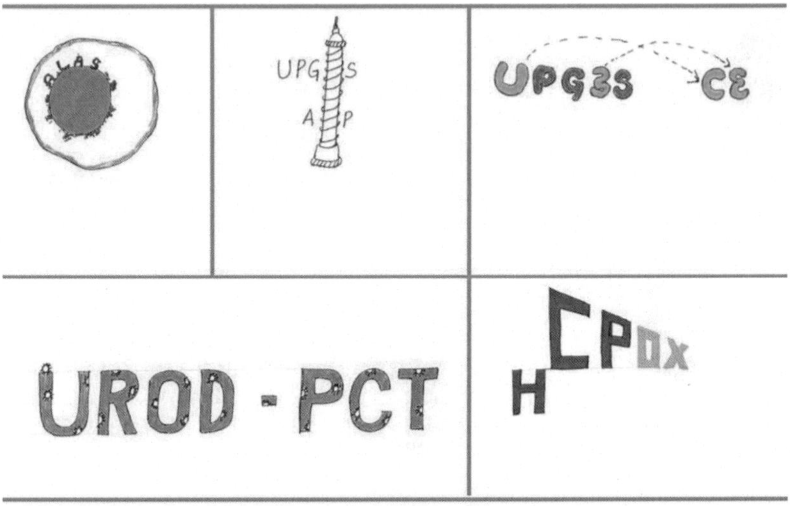

CHAPTER 10

Fill the urea cycle intermediates corresponding to the story.

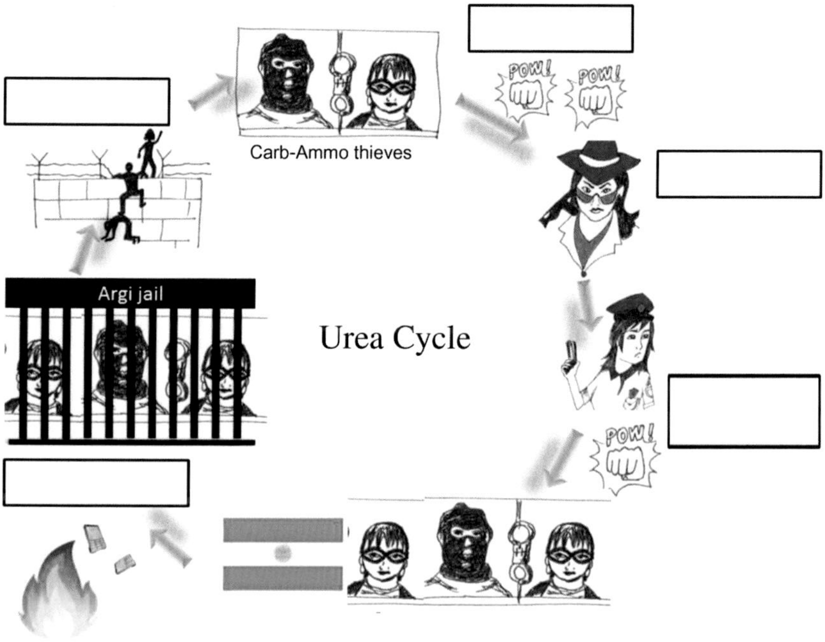

CHAPTER 11

Write the accumulating compound below the enzyme block.

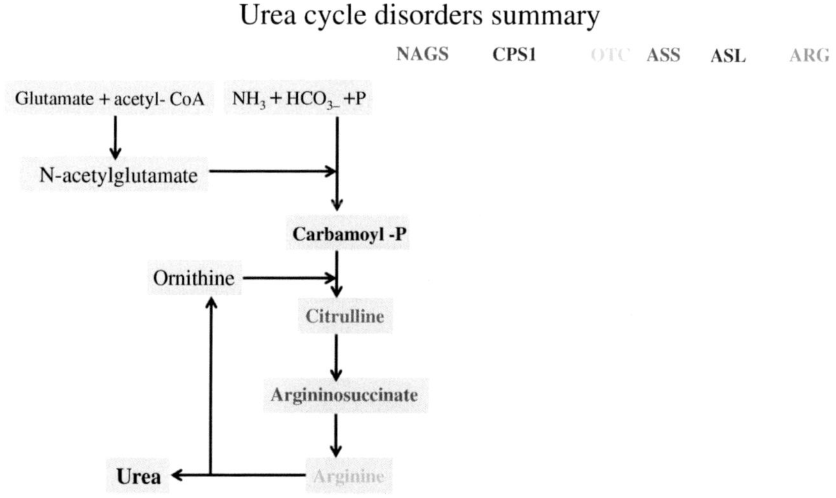

CHAPTER 12

Write the glycogen storage enzyme defects in the empty column.

Type	Story Part Correlating	Disease	Enzyme Defect
0	It's a Great Story	Glycogen Synthase defect	
1a	Von (one) Gierke was a Great 6 Pack abs warrior	Von Gierke's disease	
1b	and transported		
II	a pompe with acid maltase	Pompse's disease	
III	He adored Cori—a debrancher	Cori's disease	
IV	While he was a brancher from andery	Anderson's disease	
V	He fought with McArdle with large muscle phoscle	McArdle syndrome	
VI	to win Her as a life partner	Her's disease	
VII	On the day of Tarui, with blood (RBC) stained muscles and one PFK sword	Tarui's disease	
VIII	He suffered 8 liver powerful kuts (cuts)		
IX	And 9 muscle phoscle kuts (cuts)		
X	Finally impressed, she married him in the camp of deproka		

CHAPTER 13

Identify and write the name of the ceramide derivative.

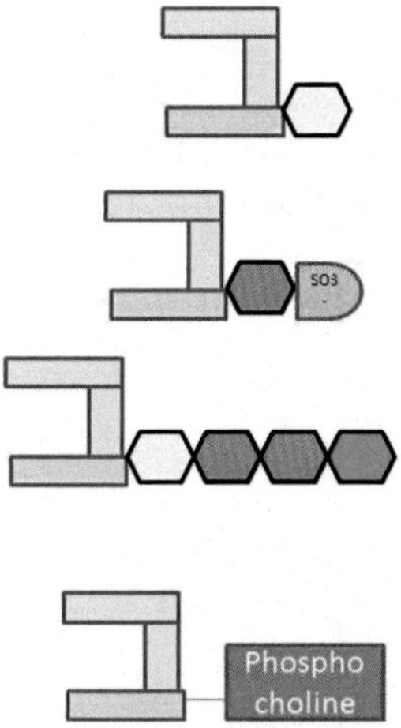

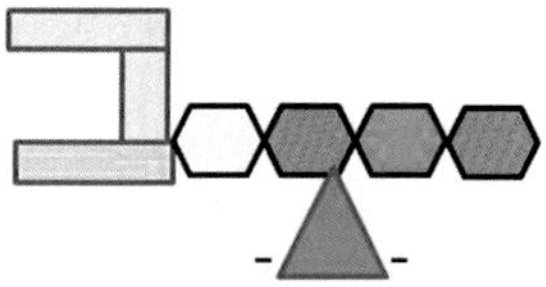

CHAPTER 14

Write the sequence of removal of the component of complex lipid.

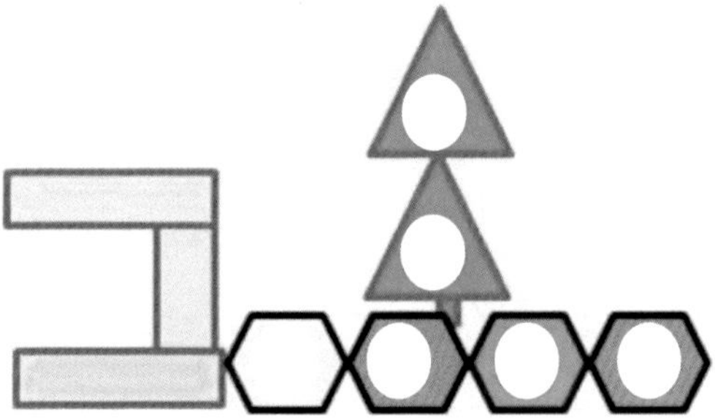

Identify the disease and write the enzyme defect.

Write the mnemonic of Tay-Sach's and Sandoff's diseases.

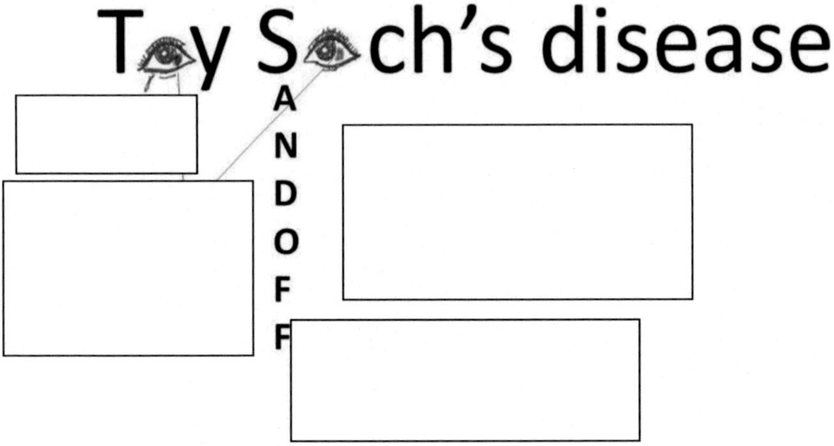

Why has this girl kept her hand on her abdomen. Identify the disease and enzyme defect.

CHAPTER 15

Identify the disease and enzyme defect. Can you write two other peoples' names who were standing with him?

Identify the defective enzyme which leads to this syndrome. Write accumulating material is the molecule which gets deposited due to defective catabolism.

Name the depositing mucopolysaccharide and the enzyme defect.

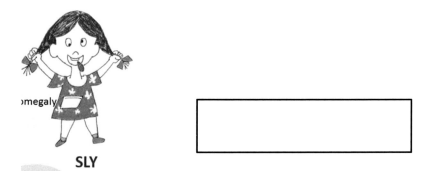

Identify the two syndromes and their enzymes defects. Write important clinical features of both.

Identify the defective enzyme of Sanfilippo syndrome and fill the empty boxes.

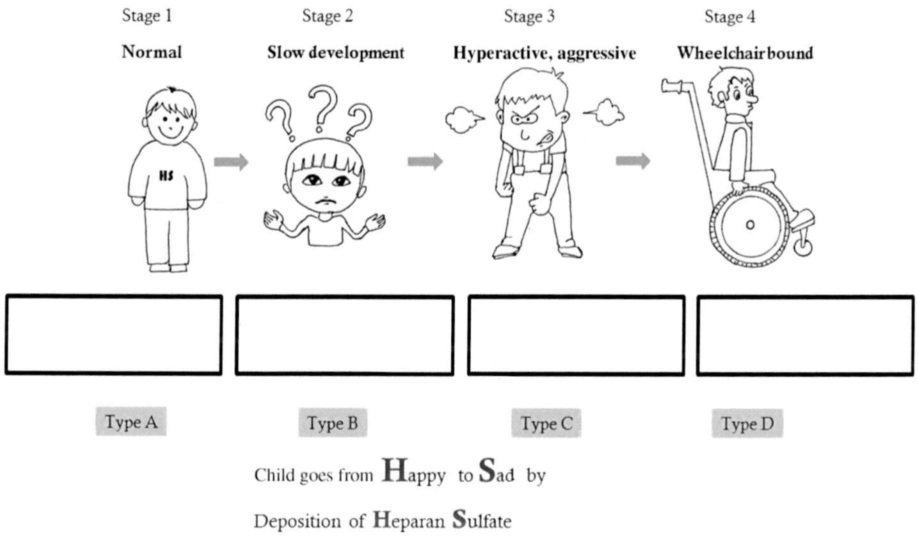

CHAPTER 16

Write the missing prostaglandins and enzymes in the spaces provided.

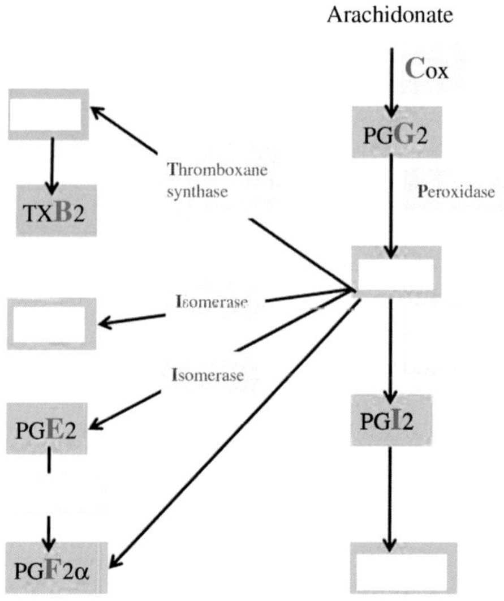

CHAPTER 17

Write the sources of atoms of the purine ring.

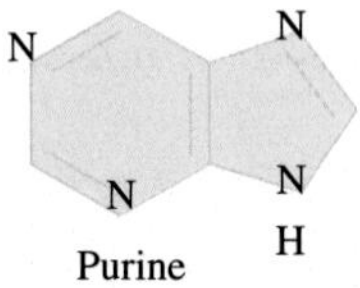

Purine

CHAPTER 18

Write the sequence of addition of purine atoms.

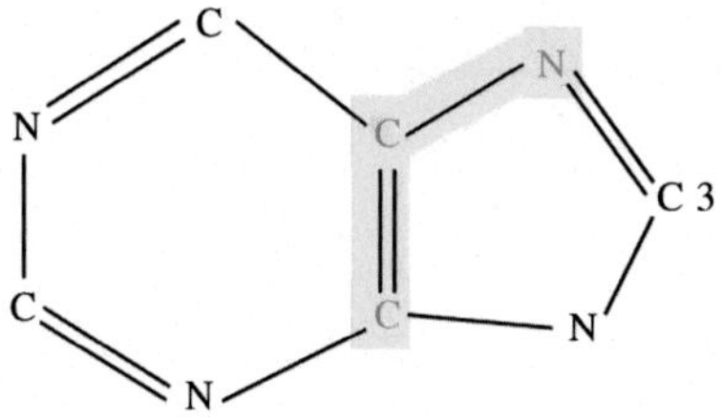

Write the names of the intermediates of the de novo pathway of purine.

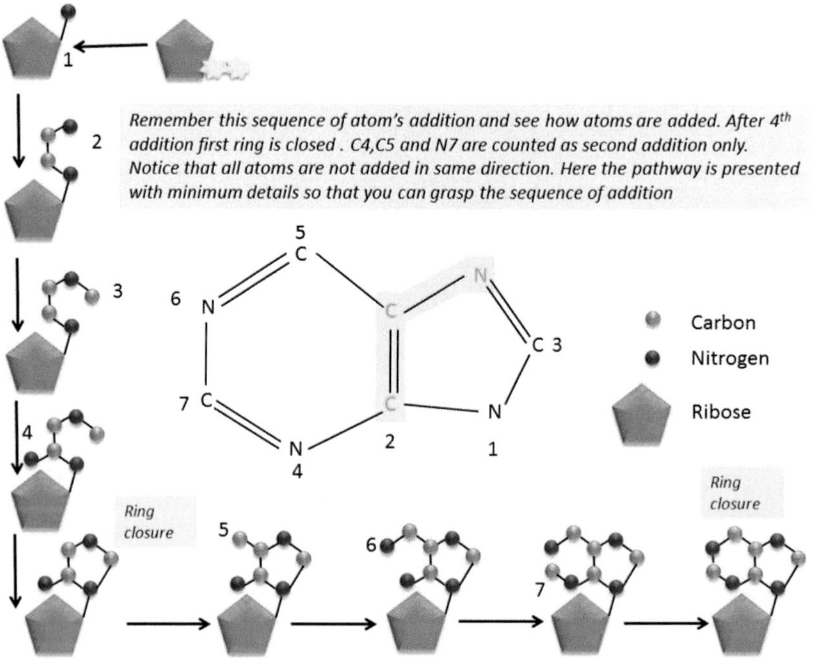
1
2
Remember this sequence of atom's addition and see how atoms are added. After 4th addition first ring is closed . C4,C5 and N7 are counted as second addition only. Notice that all atoms are not added in same direction. Here the pathway is presented with minimum details so that you can grasp the sequence of addition
3
4
5
6
7
Carbon
Nitrogen
Ribose
Ring closure
Ring closure

CHAPTER 19

Identify the three pyrimidines and write their names.

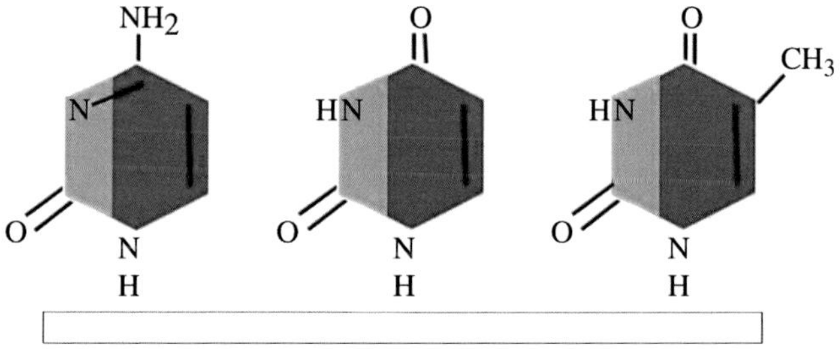
NH2
N
O
N
H
HN
O
CH3

CHAPTER 20

Write the name of the intermediates and enzymes in the boxes below.

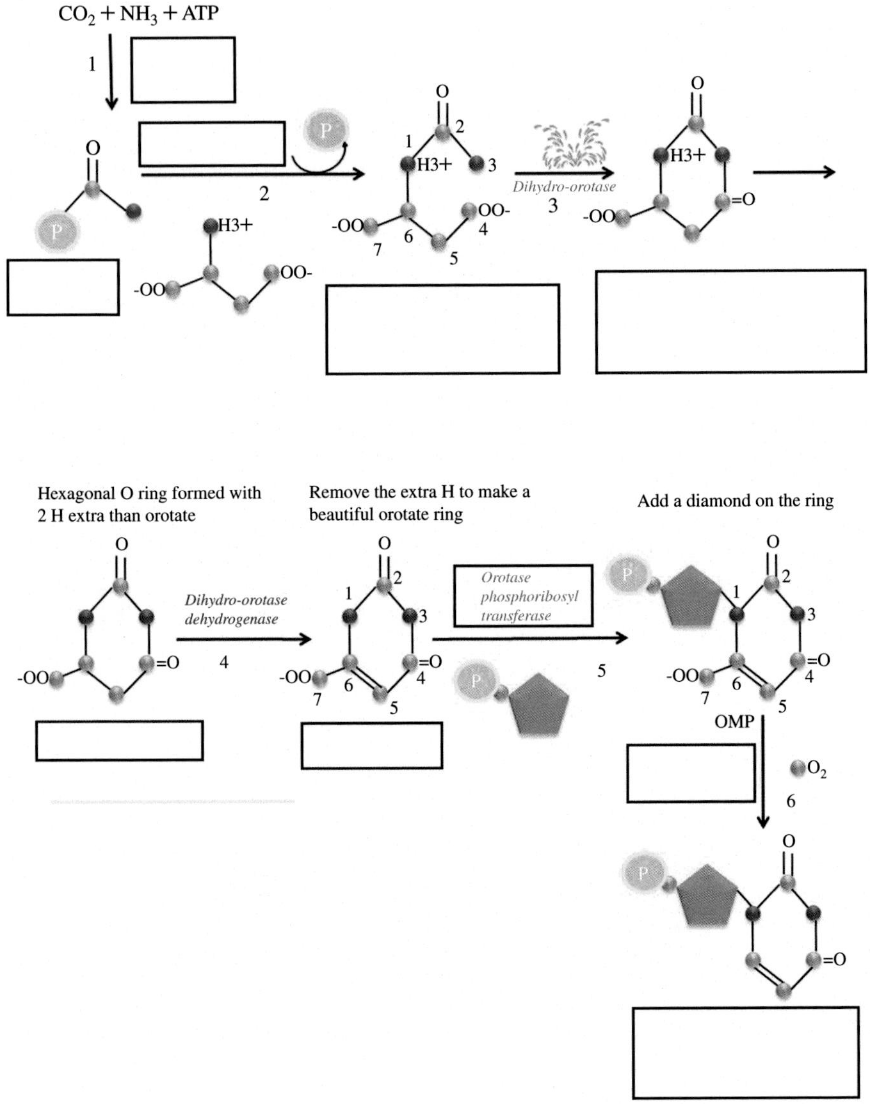

FURTHER READING

Berg, J., Tymoczko, J.L., Gatto Jr, G.J., Styer, L., 2015. Biochemistry, eighth ed. WH Freeman and Company, New York, United States.

Devlin, T.M., 2010. Textbook of Biochemistry with Clinical Correlations, seventh ed. John Wiley & Sons, New York.

Rodwell, V.W., Bender, D., Botham, K.M., Kennelly, P.J., Weil, P.A., 2014. Harpers Illustrated Biochemistry, thirtieth ed. McGraw-Hill Education, New York, United States.

Vasudevan, D.M., Sreekumari, S., Vaidyanathan, K., 2016. Textbook of Biochemistry for Medical Students, eigth ed. Jaypee Brothers Medical Publishers, New Delhi.

Printed in the United States
By Bookmasters